D1122031

T
21
G65
1986

Go~~h~~er, Michael.

Reinventing
 technology

$13.95

72795

REINVENTING TECHNOLOGY

The Institute for Policy Studies, founded in 1963, is a transnational center for research, education, and social invention. IPS sponsors critical study of U.S. policy, and proposes alternative strategies and visions. Programs focus on national security, foreign policy, human rights, the international economic order, domestic affairs, and knowledge and politics.

The Institute for Policy Studies is a non-partisan research institute. The views expressed in this study are those of the author.

Alternative Policies for America

Series editor: Chester Hartman

REINVENTING TECHNOLOGY

POLICIES FOR DEMOCRATIC VALUES

Michael Goldhaber

The Institute for Policy Studies

Routledge & Kegan Paul
New York and London

I dedicate this book to
the memory of Rose Kohn Goldsen
(1917–1985)

First published in 1986
by Routledge & Kegan Paul plc

Published in the USA by
Routledge & Kegan Paul Inc.
29 West 35th Street, New York, NY 10001

11 New Fetter Lane, London EC4P 4EE

Phototypeset in Linotron Times 11/12pt
by Input Typesetting Ltd, London
and printed in Great Britain
by Billing & Sons Ltd, Worcester

Library of Congress Cataloging in Publication Data

Goldhaber, Michael

   Reinventing technology
   (Alternative Policies for America)
   Bibliography: p.
   Includes index.
   1. Technology and state—United States.   I. Title.
II. Series.
T21.G65   1986   303.4'83   85–25735

ISBN 0–7102–0906–1 (C)
     0–7102–0907–X (P)

British Library CIP Data also available

Contents

72795
T
21
G65
1986

Series editor's preface

The "Alternative Policies for America" series offers a concrete set of programs for dealing with the country's most pressing domestic problems. It answers, in very specific terms, what radical analysts of the failure of the US economic and social system can offer in the way of alternatives. We seek to make clear that there are ways of ensuring that all Americans can have a decent standard of material well-being and public services, a healthy and safe environment, and real participation in the political processes of governance and how goods and services are provided. The problem, this series attempts to show, is not the inability to create workable alternatives, but our system's unwillingness to effect the basic economic, social and political changes in the structure of American society that will permit such alternatives to take root. Books in the series will treat housing, food, technology, health, transportation, the environment, education, energy, bureaucracy, and democratic participation.

The Institute for Policy Studies sees this intellectual work as a first and necessary step in building a movement for political change. The analyses and ideas put forward in these books must take concrete form as demands for change at the federal, state and community levels. In many cases, model legislation embodying the proposals put forth here is being developed for introduction into the policy debate, and as a means of public education and constituency building. The many individuals and groups who have been involved in the discussions and review which led to these books provide the nucleus for a continually expanding circle of activists who, we hope, will some day turn these ideas and proposals into a more just and human America than we see today.

Chester Hartman
Institute for Policy Studies

Author's foreword

In 1982, Rustum Roy suggested to me that IPS's new Alternative Program for America should include a discussion of technology policy, and that I should undertake to organize it. Chester Hartman and Roger Wilkins were receptive to the idea, and I began, slowly at first, to seek out prospective members of a panel. As with the other branches of the Alternative Program, the original plan for the technology program was that it would be a joint creation of the panel. That presupposed both a fairly sizeable budget and a considerable time commitment on the part of the panel members. We had difficulties obtaining enough of either of these. That was especially problematic for this particular subject since the democratic left has too limited a history of involvement in many areas of technology policy to have developed a set of varying positions from which a panel could have drawn together a consistent and coherent program. We therefore modified our plans.

Although I did engage in a preliminary round of discussions in fall, 1983, with some of the people I was considering for the panel, I took on most of the responsibility of drafting a program, merging that task with preparing the volume that was originally conceived as the vehicle for publishing the completed program.

I soon realized that this work allowed me to weave together a number of separate preoccupations that I have carried around for almost as long as I can remember. The work you have in your hands has thus ended up being far more personal than I originally foresaw. A very brief autobiographical sketch may help explain the motivations of what follows.

I was born, during World War II, into a family of refugee nuclear physicists who were close to, although not officially involved in, the atomic bomb project. What I was told about the Holocaust, as well as the bomb, made an early and deep impression. As a child, I was fascinated with invention, with moral issues, and with the problem of deciding what to learn. Among the family friends who influenced me were several of the founders (too late) of the scientists' movement for the control or banning of nuclear weapons. By the time I was supposed to choose a college major, I had decided that negative technological implications were too likely and unforestallable to make the pursuit of technology acceptable. Instead, I chose (somewhat unoriginally) physics, with the intent of pursuing theoretical inquiries that I was sure had and would continue to have no applications.

By the time I was well along in graduate school, I began to realize that even the "purest" of science was supported largely for its indirect value to the Pentagon. Physicists, if not their physics, were certainly implicated in the Vietnam War, in ways that seemed remarkably insensitive to the suffering they were provoking. Social injustice also did not seem to be mitigated by technology. I was not alone in such views, of course. By the end of the 1960s, I was among the founders of Scientists and Engineers for Social and Political Action.

In the early 1970s, I served on the Council of the Federation of American Scientists, which I felt deliberately ignored obvious conclusions about weapons research that would have had negative effects on members' pocketbooks. The experience helped propel me towards leaving physics. I began a somewhat difficult search for a political viewpoint that I could live with, which meant it would have to make sense, and to seem both usable and humane.

The search eventually developed into a number of intellectual projects, still in progress, from which I have drawn freely for this book. These include attempts to better understand: the origins and conclusions of wars in this century; the connections between science; technology and other aspects of modernity; and the human meaning of the information revolution. Underlying all of these is the utopian goal of reshaping social institutions, including science and technology, to better serve human needs, democratically expressed.

The lengthy genesis of this work has left me indebted to many people. A proper acknowledgement would require a full-length memoir. I regret that, due to my slow development of these ideas, some of the people I most want to thank are no longer alive to read these words. I thank each of the people listed below for one or more – in quite a few cases, all – of the following: generosity with insights, ideas, and information; valuable criticism of earlier versions of the ideas in the book; wrongheadedness so infuriating it inspired me; cherished words and actions of encouragement and friendship; and kind and timely financial support: Stanley Aronowitz, Tim Athan, Delmar Barker, Stanley Bashkin, Maurice Bazin, Ian Benson, Herbert Bernstein, Martin Carnoy, Susan Chambers, Dennis Chamot, Harry Chotiner, Frank Collins, Marcy Darnovsky, Larry Deressa, Kenneth Donow, Frank Empsak, Barbara Epstein, Deborah Estrin, Sammy Evans, Ken Feldman, Lee Felsenstein, Anna Ferro-Luzzi, W. H. and Carol Bernstein Ferry, Theresa Funicello, Medard Gabel, Giovanni Gallavotti, my parents – Gertrude and Maurice Goldhaber, Judith Goldhaber, Lucy Gorham, Allen Graubard, Dick Greenwood, Judith Gregory, Heidi Hartmann, Dick Harwood, Larry Hirschhorn, Douglas Hofstadter, Laura Hofstadter, Kathy Johnson, John Judis, Mario Kamenetsky, Merry Kassoy, Evelyn Fox Keller, Peter Kenmore, George Kohl, Frank Kramer, Mary Laucks, Will Lepkowski, Hormoz and Leila Mahmoud, Seymour Melman, Jerry Morgan, Vincent Mosco, David Noble, Kristen Nygaard, the late Harriet Older, Peter Oppenheimer, the late Frank Prager, Herta Prager, Fred Reines, Mindy Reiser, Robert Rodale, Leonard Rodberg, Bernard Roth, Rustum Roy, Tom Sanzillo, Barbara Schild,

Marion Schild, Ilse and Karl Schrag, Charles and Sylvia Schwartz, Harley Shaiken, Larry Sirott, Jeremy Stone, the late Gertrud Weiss Szilard, Paul Trachtman, Pravin Variya, Don Wharton, Patrick Esmond White, Carol and Terry Winograd, and Joel Yudken.

In addition, I would like to thank my teachers, colleagues, and students at Harvard University, Stanford University, Rockefeller University, the University of Arizona, the University of California at Berkeley, the Cornell in Washington Program, the East Bay Socialist School, and the Washington School of IPS, colleagues in various political movements, including Science for the People, the Federation of American Scientists, and the New American Movement, as well as participants and audiences at seminars, lectures, and conferences, particularly the conference on Technology and Meaningful Work which I organized for IPS and the Rodale Press in spring, 1984, in Kutztown, Pennsylvania.

I feel a special gratitude to my present and former colleagues at the Institute for Policy Studies, who together make it a uniquely excellent place to produce a work such as this. John Cavanagh, Barbara Ehrenreich, Rachel Fershko, David Leech, Nancy Lewis, Shoon Murray, Marcus Raskin, and Rustum Roy were particularly important. Robert Borosage, our director, was generous with precious funds.

I also thank: Amy Stanley and Anne Posthuma who assisted in early stages of this work; Robert Krinsky who volunteered his time to help with the bibliography; Yvonne de Cuir who provided wise counsel at difficult points in the writing; the staff of Kramerbooks, around the corner, who indulged my habit of browsing to distract myself from thoughts struggling to be prose; Stratford Caldecott who, as the US representative of Routledge & Kegan Paul was wonderfully patient; and Chester Hartman for his great skill and discretion as editor.

Finally, thanks and love to Barbara Benzwi who managed to be supportive in essential ways despite my foolishness at

undertaking this project to coincide with her exhausting
year of medical internship.

Washington, DC
September, 1985

PART ONE

Technology now

PART ONE

Technology now

1 Introduction

In the 1980s technology has become a mainstay of political rhetoric, right along with God, the flag, and the family. The 1984 presidential election illustrated this well. Senator Gary Hart's campaign was notable for its identification with the "sunrise industries" of high tech. (He was one of a band of what were called "Atari Democrats," until the Atari Computer subsidiary of the Warner Corporation, en route to total collapse, transferred its production jobs overseas.) The more centrist Democratic banker, Felix Rohatyn, while less of an enthusiast of high technology, was just as clear that the country needs to concentrate on an essentially technological task: increasing our productivity. And Ronald Reagan and his supporters were, if possible, even more enthusiastic than Hart about high technology – from the "Star Wars" approach to saving us from nuclear attack to electronic mail – and more unabashed about any negative consequences it might entail. Mr Reagan seemed not to have forgotten the television message he delivered weekly some two decades earlier: 'Remember – at General Electric, Progress is Our Most Important Product.'

What is so evident on the level of presidential politics mirrors trends in Congress, in state houses, in the mass media, among big and small businesses, the military, labor unions, and in foreign capitals as well. Virtually every state has plans to develop its own version of California's microchip center – 'Silicon Valley;' major newspapers have eagerly explored, in front page stories, such highly technical developments as Japan's 'Fifth Generation Computer' project. French President François Mitterand's Socialist government is following, and outdoing, his rightist predecessor in enthusiasm for French involvement with high technology.

All these politicians are by no means mistaken in sensing the importance of the rapid changes in technology that are underway at present. Computers, satellites, biotechnology, lasers, new materials are being modified, put to new uses, and combined in new ways – all at a fast clip. These developments stem from the decisions of technologists and corporations as well as government. It would be reckless in the extreme for any government to ignore such trends. Technological change of the magnitude now occurring is sure to have decisive consequences for our way of life.

Yet there is something seriously amiss with the commonly held understanding of technology. For all the political concern with it, technology itself is still seen as apolitical. The sunrise industry metaphor captures that misconception well. After all, the sunrise is an astronomical occurrence beyond human control. We can open our shutters to it or ignore it, but we can't keep it from happening. The 'next generation' of technology, the metaphor implies, is also like that. We can turn our backs and get left out or run over by high tech, or we can embrace it, but it's going to come no matter what we do. In that view, technology is simply a resource to exploit, a carpet of futuristic scenarios unrolling in front of us.

The premise of this book is that the sunrise technology metaphor, while it contains a grain of truth, is basically false. Technology is a human activity. Far from being apolitical, the technology that gets developed is a direct result of political choices, choices that could be made differently. Precisely because technology is so important, the political means by which we shape it are of utmost concern. If we want to have a say in our future, and our children's, it is crucial that we do not see our choice as the false one between embracing or rejecting what is bound to happen anyway.

As in every other area of life, the choices we make depend on what we hold most central. Science and technology policies are just like housing, foreign or fiscal policies in that they are shaped by definite values, aim towards definite goals, and favor specific social groups along with specific social arrangements. It could not be otherwise. A policy can't do everything because resources are limited, and some goals are incommensurate with others. The purpose of this book is to sketch out an approach to science and technology policy consistent with a politics centered on a specific set of values.

These values, which will collectively be termed *democratic values*, include commitments: to *the fullest possible democracy*; to *equal respect and decent treatment for each person*; to *as thoroughgoing as possible an equality of opportunity*; to *support for a rich community life*; to *peaceful and respectful international relations*; and to *a worthy inheritance for future generations, including a sustainable natural environment*. Each of these values will be enlarged upon at various points throughout the text, especially in Chapter 7. Most Americans would support these or similar values, but they rarely are articulated in policy formulation. In practice, that means other, sometimes opposed values are often the ones that guide choices, and therefore come to dominate.

Technology policy alone is quite obviously not enough to sustain a good society. All major areas of government policy must be examined and reshaped from a similar perspective. Books such as the ones in this series – "Alternative Policies for America" – can help existing and new movements to understand the links between their different aspirations and relate these to specific policies. It is such movements that will have to press for the policies they want and continue to see that they are actually implemented.

Conversely, technology policy must necessarily be a key part of a larger program. Technology shapes what is possible economically and socially. It affects patterns of work, community life, home life, power relations, modes of thought, and much else that is of central importance to any prohuman politics. Current technology policies, taken as a whole, move away from, rather than towards, a reasonable vision of a good society. Sharply different policies are both needed and possible.

Some of the other volumes in this series will focus on health, energy, environment, transportation, and agriculture policies – all areas with inescapable technological aspects, to be taken up in those other works, and therefore largely to be ignored in this one in favor of the broad social issues mentioned above.

In Part One, I begin by exploring the nature of technology and science. Next I explain why technology issues are currently of such special import. I then examine the workings of government policies in this area, as a means to lay bare some of the underlying values as well as the interests that benefit from and often are behind these policies. The values currently being served and the policies they spawn are in fact incompatible with the basic democratic values that guide this book.

A few words of warning are in order. First, technology and science, their social impacts, and the government programs that relate to them, are far too complex and diverse to do more, in a short book, than arrive at general conclusions – conclusions which by no means can do justice to the full complexity of all the issues raised. Inevitably, I will be slighting some valuable and courageous aspects of government programs.

Second, in a work covering technology and its social impacts from a political perspective, it would be inappropriate to include details comprehensible only to initiates, and I hope I successfully avoided them without losing a sense of the substance of what is at stake.

In Part Two, I offer alternative policies framed as legislative proposals. I am under no illusions that they could presently be passed. The purpose of being so specific is rather to show that a reasoned and practicable alternative is possible, and to offer a clear idea of how it might be constructed. The intent is to inform and enliven debate: these proposals are offered as only the beginning of what obviously must be a complex and collective process of framing a democratic technology policy. The proposals are put forward so that a new movement can have something to sink its teeth into, a basis for new kinds of commonality among groups – including especially the technological community and constituencies such as labor, the women's movement and minorities.

2 Technology and political choices

In the late twentieth century, any thought of the future which fails to take account of what technological possibilities are opening up is grotesquely incomplete at best. Yet, if political life, with all its debate, division, and struggle to define alternatives, has any meaning, it is as a means of deciding what future we want and how to get there. The conclusion is clear: technology must be a major focus of political thought. But this syllogism, important as it is, is the beginning and not the end of the discussion. Politicians have come to recognize that technology is important; what still is vague is how to act on that insight. The question is not whether technology is important for politics, but rather exactly how it is political. In other words, what kind of difference can political decisions about technology make, and what kind of decisions in this realm are possible? These questions are far from academic; precisely because technological change is so swift, the wrong direction can lead to social disaster in a short time. A better chosen direction might help avert trends toward social disaster, no matter what their origin.

To begin to formulate a political approach to technology, we first have to be able to draw boundaries around this highly abstract and broadly diffuse concept: just what is and what is not technology? Second, we have to see why these distinctions matter in the life of our society.

The word "technology" has multiple meanings. One is a

systematized form of change – more formally, "techno-logical innovation." Thus, one of the aspects in which contemporary society differs from all its precursors is that there are a large group of technologists – otherwise known as engineers, computer programmers, applied scientists, etc. – whose assigned task is to invent, develop and disseminate new ways of doing things and new kinds of things. (To make matters more complicated, "technology" – or "techno-logies" – also refers directly to both these sorts of creations: it can mean either the ways of doing things or the things themselves.)

However, if it is true that at the heart of technology is the existence of a group of people whose assigned task it is to innovate, these are not the only people in society who are expected to be inventive. A creative restaurant chef, constantly turning out new dishes, and occasionally publishing the recipes in cookbooks, is innovative but is not a technologist. Compare a chef with a food technologist. The chef relies on her or his senses, on artistic taste, and on skills developed through long practice, to produce dishes for a definite clientele; the chef will often choose fresh ingredients in the market and then decide what to prepare that day; even if the chef publishes a cookbook, there is no guarantee that an ordinary cook following the same recipes will be able to match the subtleties of the master. The food technologist, on the other hand, seeks a set of perfectly standardized subdividable procedures, in which highly repeatable results are what counts; these results are to be achieved, not by masters, but by ordinary workers with a minimal degree of specialized training at particular tasks.

In order to accomplish his or her goal, the food technol-ogist will as much as possible employ completely standard-ized ingredients, precise measuring instruments, and scientific laws and formulas. While the great chef may well come to have preferred customers for whom she or he will devise special creations or go to new heights of perfection, the food technologist assumes that customers are pretty much all the same, unknown, expecting each kind of product to be exactly the same every time they buy. While a good chef will quite possibly specialize in locally grown foodstuffs cooked according to the regional cuisine, a food technologist setting up a mechanized production process – say, for a certain type of cracker – aims to minimize the special skills needed, and establish enough controls over the process that a similar factory could produce exactly the same sort of crackers in any other location.

At the political level there is an even sharper difference. What politician would emphasize the need for new gastronomic delights? Commendable as the creativity of chefs may be, their goals are not perceived of as politically serious, whereas those of the food technologist probably are. The latter's goals may vary but they might well include radically increasing the output per worker or per factory, sharply extending the shelf-life of a food product, transforming previously inedible substances into food, or chemically synthesizing a substitute for an expensive or unavailable flavoring. In other words, while from a standard political viewpoint culinary masterpieces, along with other arts, are somehow frivolous, technology is serious because it has substantial economic consequences, transcends prior limits, or significantly transforms what people do. In fact, in calling for increased support for technology, politicians often seem to take it for granted that all technology is inherently "serious" in this way. The assumption is nonsensical – as will become clear later in this chapter: technology can be as frivolous as any essentially playful activity.

It would be easy to extend the list of contrasts between artist or craftsperson and technologist, and also easy to point to areas where in practice the two vocations overlap. But that is hardly necessary; enough has been offered to make it possible to summarize the key characteristics of technology that ought to be kept in mind for political analysis:

1 A clear preponderance of items and services available for sale in our society bear the earmarks of the process of technological innovation. These include, for example: toothbrushes, videorecorders, automobiles, lightbulbs,

space shuttles, zippers, eggs, toasters, toaster waffles, digital watches, clothes hangers, plate glass, semi-automatic rifles, paperback books, epoxy glue, light beer, and nylon tents, along with such services as providing electrical power, dry cleaning, blood tests, check processing, and security patrols. In addition, the tools and machinery used to produce the products and provide the services are themselves also the outcome of technology.

2 The very existence of technology implies a special group – the technologists – who dominate the innovation process. An important implication of having this special group assigned to innovative tasks is that a comparable role is largely denied to others, including ordinary workers or consumers – except after the fact. (The innovative process as a whole goes beyond invention and development to include wide adoption of the new, and in that way everyone has some degree of veto power.)

3 As the above list of technological items suggests, in the main they are produced as standardized, interchangeable commodities, produced in a profit-making business or in a large bureaucratic organization (or both) – usually to be sold either on a market, open in principle to anyone with the cash or directly via contract to a large bureaucratic organization. The market and large bureaucracies are characteristic of "modernity," and thus, so is technology.

4 Despite numerous attempts to maintain secrecy for reasons of corporate profit or national security, techno- logical innovation is inherently open and universalizable. In this it differs sharply from innovation in premodern soci- eties. The Byzantine Empire was able for a millennium to keep the secret of "Greek fire" while using it in warfare; the United States, with far more power, was able to keep a monopoly on the far more complex atomic bomb for only four years.

The difference is far from accidental; it is significant that the first use of the word "technology" in English was to describe an early seventeen-century encyclopaedia of the methods of various crafts. Such knowledge was previously understood to be the secret of fixed guilds, of no use to anyone else. But the early modern entrepreneurs who sought the ability to enter any business, and to increase their profits by enlarging production, hiring unskilled workers, and extending markets, were naturally impatient with the guilds' limited interest in change, their informal learning processes centered on long periods of apprenticeship, and their extravagant reliance on finely honed senses and skills. The appeal of collecting and codifying existing knowledge, of relying on scientific formulas, standardized raw materials and accurate measuring instruments was precisely to break through the mystique and mystery that enveloped craft guilds and even more deeply surrounded the products of other cultures or of Nature itself.

While the full scope of the technological project was rather clearly perceived by people like Francis Bacon as early as the beginning of the seventeenth century, it was of limited success and importance until little more than a century ago. Now, however, there are few branches of commerce or kinds of service that have not been technologized. Similarly, current technology rests less and less on rules of thumb handed down from old crafts; instead its basis is increasingly laboratory findings along with scientific explanations and formulas.

Is technology autonomous?

Technology is a human activity but it often seems to be a force of nature. There are a number of reasons for this.

1 Technology as a system always seems to improve on existing technology. Thus, if past technology has led to higher productivity, future technology, it is commonly assumed, will increase productivity still further.

2 Technological progress is based on competition among firms and nations. Thus, technology as a whole seems to work very much like the arms race: we fear the Japanese will beat us to the "Fifth Generation Computer," so we

devote resources to the same end. Whatever advance we do make, whoever is behind will try to copy and build on to remain competitive. So we will have to forge ahead even faster.

3 Decisions about the future direction of technology seem to be made according to ideas in the air in the technological community, rather than by purely individual choices. Thus, for example, as it became possible to produce integrated circuits on a single silicon chip, seemingly obvious economic considerations led to producing a general purpose circuit, i.e., a digital computer processor, rather than specially designing a separate circuit for each different function. Likewise, the goal of higher speed, higher density circuits was also obvious to many in the field.

4 On the other hand, inventions of great influence come from surprising sources, including individuals working on their own (e.g., Chester Carlson, inventor of what became the Xerox machine). Since these inventions obviously cannot be anticipated, there is apparently no way for the direction of technology to be controlled.

5 Technology builds on science, and scientists are supposedly led to their discoveries by nothing other than the nature of what they already know and their experimental and observational capabilities.

6 The results of prior technology are everywhere, often are incomprehensible to the lay person and act in unexpected ways, in this way resembling natural processes or living things. For example, without special instruments, it is as impossible to take apart a digital watch to see how it works as it would be to do the same with a housefly.

7 Finally, products of technology are now more feared than are natural forces. No one anticipates a natural disaster on the order of the "nuclear winter" that might result from nuclear war. (Indeed, theorized catastrophic collisions of the earth with asteroids, such as the one that is hypothesized to have caused the extinction of the dinosaurs, could now probably be prevented by utilizing existing space technology.)

Do all these reasons hold? Is technology like a train leaving the station that we can hop onto or get left behind, but whose destination is beyond our control? If so, then it would be correct to speak of "sunrise" industries; it would be correct for politicians simply to urge improving the conditions for technology. At best, a more nuanced political approach to technology would only involve ameliorating its negative consequences.

The answer is that technology is not autonomous; the apparent relentless, natural forward motion of the field is in reality anything but that. Closer examination of the arguments advanced above (or of similar arguments elsewhere) will reveal a very different conclusion: technological progress is always guided by values and interests that come from outside technology. Let us proceed once more through the list above to see what has been left out of account.

1 The first five items concern technological innovation as a human process. Each of these points leaves unstated that there is a complex social – and therefore ultimately political – process by which technologists, scientists, industrialists, etc. decide on such questions as what constitutes an "improvement." A technological development is always planned for some social setting in which the improvements involved make sense. But what makes sense is always a matter of social consensus, in which a combination of cultural, economic, and ideological elements are socially evaluated to arrive at a set of priorities. For instance, in automobile design there are many possibilities of what might be considered improvements under different circumstances: cars could be faster, safer, use less gas per mile, be bigger to hold more passengers in greater comfort, be smaller to be more easily parked, require less maintenance, or wear out faster (planned obsolescence).

Simply put, what constitutes an improvement is always a question of values and dominant interests, and it is never a purely technological issue. Even though technologists often seek numerical measures of performance by which to gauge improvements, what they choose to try to measure depends upon non-technological factors. If pleasing working conditions were valued more highly than efficiency, for instance, it would be possible to arrive at a variety of numerically measurable quantities, which at least partially or indirectly would correlate with this concept. Technological progress could then be measured according to the new parameters.

If technologists were to operate without some guiding set of values, what would count as improvement would constantly change according to the particular measure that happened to catch their momentary attention. Even if they were to persist with some single measure, it would be more likely to be one of the huge number of socially meaningless possibilities than anything that could be called serious. For example, an auto designer might take as a criterion of perfection how close the weight of the brand new car was to exactly two tons; there would be no limit to improvement, for there would always be another decimal place of accuracy to consider. The effort might make for a real challenge to technical skills, but there would be no noticeable benefit to anyone else involved. Needless to say, no corporation or government agency would be likely to support the pursuit of such a goal – unless, mystified by technology, it thought it was doing something else. Technological frivolity of this kind does take place, but only most rarely and accidentally could it have any important consequences. Technology is significant precisely when the goals it adopts are related to values and interests; then the question becomes what values and whose interests are being served.

The range of values that can motivate technological development is wide: from authoritarianism, racism, and destruction – witness Zyklon B poison gas in the Nazi concentration camps, machine guns, and the South African automated pass system; through corporate power, profitability and hierarchy – consider the assembly line, centralized data banks, large office buildings, and containerized shipping; to human equality (at least roughly) – as in mass transit or improvements in nutrition. Encouraging curiosity, extending life, facilitating playfulness, or enlarging democracy have all been goals as well. The values that matter most are those of the institutions and individuals with the greatest power to determine the direction of technology in our society. Overall, at present, these are the values of corporations and the goals of certain government agencies (see Chapter 5). But, as is especially evident in the latter case, these values and goals can be changed by different political choices.

2 Competition among nations or even between firms is not implicit in technology, and there is no reason technology has to be shaped accordingly. What is true is that technologists have commonly used the potential for competition, and the political weight commonly accorded it, as a basis for urging support for technological development. They have been fairly successful with this tactic, and, as a result, they have been expected to deliver in terms of advantages over the competition. But it does not therefore follow that such competition is either wise or unavoidable. In fact there are many instances of technological cooperation – for example, making air travel safe – that demonstrate it is possible for nations and corporations to break free of competitive patterns when it is widely acknowledged that they are harmful. The will to do so on a larger scale is a political question.

3 The foregoing should suggest that ideas "in the air" are there for more than purely technological reasons. A bread with triple the current calorie content may be technologically feasible; indeed, it is quite possible that a development in baking technology would point in that direction. But since our culture as a whole would accord little value to such an innovation, it would never be a serious focus for technologists. Clothing that would dissolve in the rain, a pill that would mimic the effects of hayfever, or assembly-line processes so designed that the sounds emanating from them provide renditions of all nine Beethoven symphonies may all be quite feasible, and might offer interesting challenges to technologists. None of them are "in the air," because the context of power and values in which technology is

embedded makes these seem nonsensical at the same time as it finds sense in cruise missiles, space stations, and increasing productivity in an era of high unemployment.

4 Individual inventors, or even groups of inventors, can undoubtedly depart from dominant values. But for their inventions to have any significant impact they must be accepted by large corporations, venture capitalists or government bureaucracies, then by some institution involved with distribution, and finally by some group of users or consumers. An inventor like Chester Carlson may have had difficulty finding a corporation willing to invest in developing his electrostatic photocopying (Xerox) process, but there was nothing remarkably eccentric in his awareness that others besides himself frequently could make use of copies of business letters and other documents. In the 1950s, few individual inventors would have worried about problems of energy conservation; likewise, an inventor of today would probably have little motivation to work on an idea of specific benefit to welfare mothers.

5 Since technical feasibility does not in itself determine the direction of technology, it is obvious that new scientific results do not either. The next section takes a closer look at science itself; though stemming from more complex motives than technology, science too turns out not to be autonomous.

6 This and the next point concern past technological developments already in place. Admittedly, the inventions cannot be unmade, but there is no reason they need determine the future. To a large extent, for example, the unintelligibility of present technology is a deliberate choice, for reasons ranging from laziness to the wish to extract high replacement and repair charges ("no user-serviceable parts inside"). It would be possible to put more effort into making each technological project comprehensible to its users; alternatively, new products could be designed with intelligibility as a goal. For – unlike nature – the design of modern products began with conscious understanding, and that can be made accessible.

7 As far as the dangers from nuclear war or other direct technological dangers are concerned, these demonstrate the power of technology but not its inevitability. The reason that weapons are so highly developed is a consequence of the fact that destructive power is both highly valued and easier to enunciate than the more complex set of values associated with sustaining and improving human life. Technologists are complicit: many willingly help perpetuate the arms race because they benefit from it in terms of jobs and status. But it is a national political choice that has made it easier to be assured recognition and employment in weapons development than in projects related to other values.

The place of science

In policy as well as practice, science and technology are interlinked. A political approach to technology has to deal with science as well, in ways that take into account both the linkages and the important distinction* between the two forms of activity.

---

* For readers approaching this subject for the first time, the following summary may be of help.

Very loosely, the distinction between science and technology is that if the immediate aim of technology is to achieve practical ends, then the immediate aim of science is increased understanding, or the accumulation of knowledge, especially the knowledge of nature. The term "research" may – again, loosely – be taken to describe science as an activity; "development" is the same for technology. Science or research are both often further subdivided into "basic" (or "pure") and "applied." This distinction has to do with the closeness with which the knowledge likely to be gained is consciously connected to specific possible technological applications. The discovery of nuclear fission involved basic research, but once the practical possibilities were recognized, subsequent experiments to aid in the construction of the atomic bomb were applied, even though an outside observer would have had difficulty seeing much difference in the laboratory procedures involved. Although for administrative purposes, the definitions may be made sharper, basic and applied efforts in fact shade into one another. Finally, it is, of course, unwise to read any moral connotation into the term "pure."

Like technology, science is a characteristically modern kind of knowledge. Even more than technology, it is intrinsically open, in that for a scientific result to be considered valid it must be reproducible. Reproducibility implies a set of published procedures that any other skilled scientist or group of scientists can use to duplicate the experiment or observation, regardless of any personal beliefs, virtues, conditions of birth, or particular location. All that is needed is some apparatus describable in numerical terms, often built with readily available parts, and itself understandable according to scientific explanation. Within a particular explanatory and conceptual framework, two very different-looking experimental setups may be said to lead to the same result.

What counts as a scientific explanation changes as theories change, but it characteristically involves natural objects and forces that lack intention, volition, or symbolic meaning – i.e., there are no gods, spirits, angels, demons, ghosts, portents, etc. Science assumes and reinforces a desacralized world view, one that is amenable to commercial development and bureaucratic management.

Technology is necessary for science – in effect, it is the source of scientific apparatus and at least some scientific procedures; and science is necessary for technology, in that it constantly offers new reproducible situations that can be converted into industrial scale forms, and in that through scientific explanation technologists are able to understand how to approach the goals that interest them. For example, the physics of Isaac Newton provides the basis for calculating the dynamics of the orbits of communications satellites, without which they could not be launched. The more "basic" or "pure" the science, the more its explanations are likely to tie together diverse phenomena, and the wider its potential use by technologists is likely to be.

Just as technologists are paid to innovate, scientists are paid to make new discoveries. (Again, scientists are not the only discoverers; poets, psychoanalysts, and investigative reporters all make discoveries different from science.) As a community, scientists can be said to be always seeking to discover the "natural laws" that lie behind appearances; since appearances include the currently accepted laws, in effect scientists, whether they recognize it or not, are always seeking to undermine the apparent limitations on actions those current laws suggest. In asking the question, "What is there?", scientists are really asking how does such and such work, with the implicit goal of understanding how it can be made to work differently. When Newton was trying to discover the physical laws underlying planetary motions, he was implicitly asking how to go about changing those orbits. Implicit in the study of hormones affecting human sexuality is the possibility of changing the nature of that sexuality (for instance, in sex-change operations). Thus, scientific exploration suggests and promotes new technologies (perhaps only for the far future); it may seem that this happens without any specific values or interests other than curiosity, but, again, closer examination suggests that is not so.

To indicate the deep way that values enter science, I shall adapt and summarize the most throughgoing account I know of – that of the late philosopher of science Imre Lakatos – on how scientists decide between rival theories. Scientific theories can never be proved or disproved by experiment, since the connection between theory and experiment is always open to interpretation, and interpretations can always be modified or elaborated to account for any disagreements or to explain away agreements. How then are better theories selected?

Lakatos suggests that the scientific community functions like the stock market, in that scientists choose between rival explanatory systems on the basis of which ones are likely to undergo the fastest growth. That is, each explanatory system suggests some new concepts that relate to new kinds of experiments and observations; these in turn help suggest further elaborations and modifications of the explanatory system and thus lead to still further experiments. The systems in which the payoff of interesting new concepts and interesting new experiments is likely to occur fastest have an obvious attraction for scientists since these will help not only in furthering their careers but in placing them closer to heretofore undisclosed knowledge of nature and to novel technological possibilities.

The catch in this explanation is the word "interesting." What is interesting remains a human question answerable differently by different people and at different times. Since one very strong limitation of experimentation is what experiments can be funded, and what social purposes are likely to command the development of new technologies, what is scientifically interesting, and therefore the character of the explanatory system that is likely to survive, will be influenced by who has political or economic power and to what ends that power is exercised.

The choice of explanatory system in which to pursue knowledge influences what experiments are done, and how they are to be interpreted. Thus, the very nature of scientific knowledge and the set of known facts depends in a very complex way on the power structure and values of the society that is – in the final analysis – doing the asking. Since the choice among explanatory systems appears to be a decision about what is true and what is false, truth turns out to be highly, if indirectly, dependent on the larger society of the day.

Furthermore, it is of course not coincidental that the model of science turns out to resemble the stock market. Ideologies of economic growth and of the growth of scientific knowledge evolved at the same time and continually reinforced each other. Rather than splitting apart, these two systems are actually tending towards each other, as exemplified in the recent introduction of genetic engineering stocks on the market. Profitability and scientific truth easily become intertwined. Just as, at the leading edge of technology, companies compete with one another to be the first to produce a new type of commodity (e.g., the first "256,000 bit random access memory chip" – better known as the "256K RAM" – or the first human insulin genetically engineered into bacteria), so scientists compete with each other to be the first to discover something anticipated by the current explanatory system (e.g., the first hormones isolated in mammalian brains or the "top" quark). The sheer joy of being first is common to both such enterprises; so is the increased likelihood of obtaining not only recognition but funding as a result of demonstrating speed that can be related to fast growth – in the one case of profits, and in the other of the explanatory system itself.

The overall direction of science and technology

So far in this chapter, technology and science might each appear to be a single seamless unit. The reality of course is far more complex. Some technological efforts focus on highly specific problems such as how to decontaminate the Three Mile Island reactor. The values and interests involved are usually easy to discern. But these specific projects normally make use of other technologies that are less specific in scope. In the Three Mile Island case these would include technologies ranging from structural engineering to video transmission to chelate chemistry to radiation detection, among many others. Such technologies in turn involve others, such as the metallurgy of steel for the structure of a crane, semiconductor electronics for a video camera, and so on. These more general purpose technologies are in turn closely connected to applied science; at an even further remove, various basic sciences would be involved.

The further we go from specific applications, the more difficult it usually is to ferret out the values and interests that underlie the activity. If we look at some particular laboratory, or at one person, or even one special field, the discernible motivations may be quite idiosyncratic. There may well be an engineer somewhere working on improving the color resolution of television screens because he or she would like to see the face of a certain performer better; there may be chemists who just love the smells associated with working on certain compounds; there are computer scientists convinced that computers will help make all people equal. These individual motivations are not to be dismissed as necessarily irrelevant, but they also do not normally determine the overall direction of any particular field, much less the direction of science and technology as a whole.

Roughly speaking, each different subspeciality in science and technology may be viewed as a service to all the other subspecialities where it is applied. It takes on the sum total of the values and interests they serve, approximately in proportion to the degree of demand each application places on the particular subspeciality in question. At each level in this process we must include the institutional interests of the members of the subspeciality and any related bureaucracy. Especially for the sciences, we must also include what might be termed ideological applications – thus the field of ecology may serve the values of the environmental movement, and the field of sociobiology to some extent serves to support sexist ideas.

As each field then builds on past foundations, it continues in directions suggested by the values and interests it has been serving – that is, it expands more in directions helpful to those values than in other directions. In ways both subtle and direct, these values come to imbue the thinking of members of the subspeciality: thus, the value of increasing efficiency and raising productivity becomes central to the thinking of industrial engineers; computer scientists working with them will have these same values reinforced, and so on. The values and interests will also be embodied in the procedures, processes, and product designs emanating from each subspeciality.

When a technology or a science now is applied in a new way, the values and interests that have shaped it before will influence its suitability for the new application, and may even limit the ways the values and interests directly related to that application are served. For example, for most of their four-decade history, computer systems were developed to serve the needs of large, more or less bureaucratic organizations. Now the programs available for personal computers are being written as if the individual user will operate as a scaled-down version of a bureaucracy (with programs for "database management," "word processing," and balance sheets). This emphasis influences not only who can use such systems but how the users will come to view themselves. This is a particularly clear-cut and easy to see example; many others, from the most basic science to even more directly applied technologies that illustrate this same point may be found in the chapters that follow.

Technology and science together both amplify and help perpetuate dominant values. To the extent that these values accord with the broad interests of humanity as a whole, with the needs of the downtrodden and ill-served, that is all to the good. Unfortunately, the motives of profit, international competitiveness, national expansion, and perpetuating the power of the already powerful tend to prevail in the institutions that currently set the direction of innovation. Deep changes in these institutions are sorely needed.

The basis for change

Despite the fact that technology and science have been and are being overly shaped by corporate and militaristic values, it is never too late to change – barring all-out nuclear war. The overall direction and the direction of specific sub-fields may be shifted. Technologies already developed may be converted to other purposes, but this will not be as easy as if they had been built up from the start according to more human-centered values. The larger the range of applications of a particular technology, the more feasible it will be to turn it in other directions. But there is a need for continuing awareness that existing technologies might prove to be ill-suited to new purposes. In addition, the overall mix of technologies and sciences might well require significant change.

Since current trends are both accelerating the rate of

technological innovation and worsening the values embedded in them, it is doubly urgent that we begin changing direction as soon as possible.

What technology cannot do

Technology is molded by dominant values and interests but it is not capable of delivering exactly the results that that influence would suggest. There are three reasons:

First, non-dominant values do shape technology to some degree; for instance, there were radical democratic impulses behind some developments in fields such as solar power and personal computers. In both cases, a crucial period in which advances were possible with small-scale efforts happened to coincide with the breakup of the sixties' peace movement and counter-culture. People whose values had been formed in those movements became the "ecofreaks" and "hackers" who partially founded the new technologies.

Second, dominant values are not completely consistent with one another; the inconsistencies, in technological form, open up possibilities of moving in other directions. Thus, automation technologies are normally backed as a means of raising productivity, thereby improving management's capacity to control the supply, and thereby the price, of labor; but they also raise the question of why anyone should have to work for a living – a question which, if taken seriously, might move the technology in very different directions.

Third, the basic nature of technology taken as a whole is at odds with some of the aims frequently assigned to it. Perhaps its most important aspect in this regard is its "universalizability". As was stated at the beginning of this chapter, technology is intentionally, and unavoidably, an open system of knowledge – in opposition to secrecy and exclusive, unsharable methods that typified the activities of stable and tightly knit groups. A family, a peasant village, a craft guild or a tribe can maintain secrecy or can make innovations that simply do not come to the attention of the outside world. But successful technologies are inherently based on standardized systems of operation, on publicly accessible knowledge, and even, ultimately, on readily available raw materials.

While it is possible in principle to keep the very existence of an innovation secret, no modern country or company is likely to see any benefit in such total secrecy. For example, a country that had a weapon so secret that no other nation was aware of it could not use it either to deter attack or to get its way by direct or indirect threat. Enough information would have to be released to make the existence of the weapon seem plausible. A Soviet announcement that they could now turn the entire Reagan Cabinet into frogs would have no force. It would be literally incredible, as the necessary scientific and technological breakthroughs that would have to have been involved would be too momentous to make it at all conceivable that they could be confined to any one country.

By the same token, a company with a new, highly efficient production process would certainly want to use that process to make inroads against its competition, even though that would probably give the existence of the process away. Knowing that a particular innovation is possible is an enormously important – and time-saving – first step towards figuring out how to achieve the same result, even if the means differ widely in detail. Since technological capabilities as a whole continue to grow in number, a competitor starting out later has a dual advantage in knowing what results are possible, and in having more and – often – more powerful means to achieve them. Thus strong competitors can often catch up with or even outdo the original leader in a particular technological innovation.

Consider what this means for current international "races" such as the arms race or the race to increase industrial productivity. In the arms race, destructive power keeps increasing, but neither side can gain a permanent edge. In fact, each side's temporary advantages are soon offset by an equal development on the other side: thus the fission – or atomic – bomb, first developed by the US, was offset four years later by the Soviet atomic bomb; the Soviet fusion – or thermonuclear – bomb was offset by a comparable American bomb a year later. With slightly different time-tables, the same has been true for bombers, ballistic missiles, submarine-launched missiles, cruise missiles, etc. The current US "Star Wars" effort (or Strategic Defense Initiative) is still years away from any kind of success, yet it has already surely instigated some parallel Soviet project; equally certainly, it has set in motion a Soviet study of possible offensive countermeasures, and the latter must have an American parallel as well. Even though much of the necessary technology and applied and pure science for such projects is developed in "secret," these efforts are too much part of an inherently open and international system for the details that do remain secret to matter much. Universalizability of technological development means that everyone loses in a competition in which the aim is to take power from the other side.

The ironic results of arms races come into sharp focus when we think of the origins of nuclear weapons. Much of the World War II Manhattan Project was motivated by a fear that Hitler's Germany would be the first to obtain these weapons. After the defeat of the Nazis, it became evident that the German bomb project had been operating at too small a scale to have any chance of success. Yet in the ensuing forty years, nuclear weapons have proliferated; there is no way to keep them from the hands of some future Hitler, and it is not inconceivable that such a figure could eventually take power in the US itself.

The problem of competition in which both sides lose is not confined to an arms race. Current American efforts to increase economic competitiveness by increasing productivity are sure to lead to parallel increases by industrial rival nations such as Japan or France. If productivity grows without efforts to increase overall demand for goods and services by spreading buying power, the result can only be increased levels of world-wide oversupplies, leading to heightened world unemployment and increased inequality between rich and poor. In addition, without careful restraints, increased productivity can lead to accelerated environmental destruction – as, for instance, when faster means of cutting timber lead to lower world prices, resulting in efforts by each supplier nation to sell off ever more stocks in order to maintain total income.

From many standpoints, then, efforts to use technology to gain advantages over moderately capable competitors eventually backfire; the competitors quickly are on a nearly equal footing and the overall level of competition has increased. Thus the "Golden Rule" of technology should be "Do not develop anything you would not want your worst enemy to have" – its "Categorical Imperative," "Develop only that which will contribute to the general good when everyone has it." A very large part – most likely a substantial majority – of current technological development ignores these rules. A sound technology policy would emphasize them.

Attempts at secrecy, far from being an effective alternative, only make matters worse. Opponents who are in a position to do so simply increase their independent proficiency so as to be able to keep up. The more technology and science enlarge their scope, the less substantial the secrets that can be kept. Current Pentagon efforts to prevent technology transfer to the Soviet Union are no exception. They merely mislead the American public into believing that if only we can keep our secrets, victories over the Russians are inevitable – a completely erroneous conclusion. Just as in the case of the "secret" bombing of Cambodia – clearly no secret to those being bombed – it is ordinary citizens who are really being kept in the dark by their own government or by industries that claim to be serving them.

Meanwhile, efforts at secrecy serve to alert people who have access to some of the numerous secrets that they have a potentially valuable and highly saleable commodity at their disposal. Recent espionage cases have rarely involved ideological loyalty to the other side; rather it is simply a matter of an irresistible commercial opportunity. As long as buying technological plans is cheaper than repeating a development process, all sides are sure to resort to this expedient whenever possible. (Ironically, the more loyal an American is to the concept of free enterprise and "going for it," the more she or he will be tempted to sell national or corporate secrets should the opportunity arise.) Efforts at secrecy surrounding a particular development project should suggest that the development probably violates the true interests of the public; if the attempt at secrecy is to prevent a potential enemy from benefiting, it would be better to follow the Technological Golden Rule just set forth; if no enemy would benefit, then why the secrecy?

Technological leadership

If technology is universalizable, it can be practiced by any country. But technology is also divisible into any number of specialities. The result is that as the total world pool of technological activity grows, it becomes impossible for any one country to maintain a lead in all fields. The US spends roughly half of the world's recorded research and development funds; this probably amounts to slightly less than half of all technological activity, since Soviet figures are unknown and since some must go unrecorded. It is easy for other countries – even technologically "small" ones – to concentrate their efforts in directions different from those we choose. As a result, America is inevitably losing ground as the undisputed leader of all advanced technologies. No matter what we do, that trend is likely to continue, and in itself is far from disastrous.

But we could help reduce any possible harmful consequences to the US by devoting less of our resources to self-defeating activities such as weapons development; we could further offer our capabilities to the world to help foster genuine cooperation. Such measures would make sense. To the extent that we feel compelled to adopt technologies originating elsewhere, we must understand that we thereby will come under some influence of the dominant values of those other places, just as our technology has spread our values. A reasonable response would be to see to it that our technologies more fully reflect the values we most want to preserve. What is impossible is to reverse the trend by trying to maintain a lead in every conceivable area, as some critics of current policies seem to be calling for.

Technology as pseudogovernment

The preceding section emphasized competition among countries, implying that governments play a key role in technology – as indeed they do. This fact alone makes technology a fit subject for political concern. But it might be objected that much technology is developed by and for the private sector – which is also true. Even if we ignore the many ways in which governments provide the underpinnings for private sector technology, there is an important way in which that technology is intrinsically political – namely, in its operation as a "pseudogovernment."

The pseudogovernment, like real governments, operates on various levels. At the lowest level, technology implies standardization – uniform, repeatable ways of doing things, utilizing standard objects and standard units. With technological innovation come new units, new objects, and new methods – again, all standardized. All these standards do not simply pop into being; they are deliberate – if partially arbitrary – choices made at times through a simple process of economic competition, but more often through a process of formal or informal consensus-building and negotiation among technologists and scientists, or sometimes among their employers. These standards, whether they refer to automobile engine oil performance, screw types, the numbers of pins on microchips, or the characteristics of video cassettes, spread through the world with the technologies they are part of. They channel and focus human actions with the same fixity and force as laws established by sovereign governments – such as those determining speed limits, drinking ages, or time zones. An example is the standard computer code for letters of the alphabet; lack of provision for diacritical marks makes work in most European languages cumbersome. Although individually not of great significance, they occur in such numbers that they obviously come to shape social life in very substantial ways.

Governments, of course, do not limit their actions to these seemingly minor, everyday matters, and neither is technology limited to that level of impact. Some of the more noteworthy and all-encompassing issues in which technological choices are as marked as any government action will be detailed in the next two chapters.

If the technological pseudogovernment is left outside the arena of political debate, then political activity can only become increasingly irrelevant as a means of controlling what shapes our lives. A responsible technology policy must recognize the importance of democratic involvement in the decisions now made – sometimes with little conscious reflection – by this pseudogovernment of technology.

3 The social impact of technology

The preceding chapter indicated how dominant values and interests shape the direction of technology. That direction matters because as new technologies come into being they open up possibilities that substantially alter the realities of our lives. Over the last two decades, the environmental, health, and safety effects of technology have deservedly been the focus of considerable public attention – partially, no doubt, because these effects are themselves amenable to scientific observation. There has been much less concern for the political, economic, cultural, and social consequences of technology. To be able to rethink technology policy – the ultimate aim of this book – requires a clear sense of how the major societal impacts relate to values such as equality, democracy, and minority rights. That is the purpose of this chapter.

There are many ways to conceptualize or categorize the social impacts of technology. A common procedure is simply to examine specific impacts of particular technologies, such as the effect of television on ways of perceiving violence, or the effects of new technologies on employment within a specific industry, or even in a specific plant. Such studies are the raw material of any more general overview, but more of an overview is needed to understand the consequences of technological decision-making as a whole. The approach of this chapter is to focus on some key aspects of a society and to show how technological decisions can affect them. These broad categories include wealth, power, cultural patterns, gender relationships, community, geography, and work. These categories are neither complete nor mutually exclusive; they have been chosen because they seem to include many of the most important ways we have of thinking about a society. Furthermore, their relationship to technology is often apparently overlooked. Explaining these relationships takes up most of the chapter.

The last part of the chapter relates more fundamental changes in technology to broader economic and social transformations. That discussion then leads into the next chapter, which focuses on the current emergence of high technology and on major social changes it is likely to precipitate if existing policies are not sharply modified.

Technology and the redistribution of wealth

Technological innovation is one of the main means of creating new wealth. In our society, wealth has two aspects. It indicates a level of potential well-being, comfort, or "quality of life"; it also measures the capacity to obtain what one wants on the market. The two aspects are not completely interchangeable. A person might have comforts that would be envied in other societies – say, a car and a television set – and still might not have the wherewithal to afford enough food or shelter to guarantee survival. If the car and TV are old, even selling them to obtain cash may not be possible. Thus the pattern of wealth differences within society is perhaps an even more vital question than its overall level, and technological innovation usually affects both. Therefore a call for new technology is a call for the redistribution of wealth, which, it is useful to bear in mind, can go either way – from rich to poor or from poor to rich.

Examples abound. A means to exploit a previously unusably low grade of ore transfers wealth away from regions with high-grade deposits towards those with lower grades. Such a change generally lowers the total price of the mineral and thus also transfers wealth towards its users. The invention of steam power lowered the relative worth of land in locations suitable for water mills. New forms of transportation, such as railroads, increase land values in previous backwaters. In the late nineteenth century, elevators increased already high land values in the downtowns of major cities. Air conditioning has opened up some areas with hot and humid climates, including much of the "Sunbelt," to modern commerce and industry. Economic modernization of the American South, in turn, probably helped stimulate the civil rights movement there. Mechanical and chemical revolutions in agriculture helped undercut the markets of traditional farmers, causing them to go bankrupt by the millions.

Skills often lose their value when technology introduces alternative means to produce their products. Whole valleys in Switzerland have declined as a result of the transition to electronic, digital watches. Everywhere, watch repairing is a skill of much less value than it was in the days of mechanical movements. On the other hand, mathematical skills have increased in value with the advent of computers.

To the extent that wealth is associated with comfort, or even with status, electrification helped equalize wealth by enabling the poor as well as the rich to have such conveniences as light at night and clean clothes without the drudgery of washing by hand. Private planes improved the comforts of the rich, but the noise and air congestion the planes create is a dead loss for almost everyone else.

A final important aspect of wealth is the related concept of ownership or property. Different cultures do not necessarily agree on what can be owned or on what kinds of possessions may be exchanged on any sort of market. (Ancient Athenian citizens could own land and slaves, but could sell only the slaves; contemporary Americans can neither own nor sell slaves, whereas we can both own and sell land.) These disagreements are a matter of convention or law and not technology, but what technology makes possible permits certain laws to be enforced and not others. The fact that

land is sellable today is undoubtedly dependent on the existence of surveying technologies that allow lots to be identified unambiguously in records of land transactions. Similarly, one reason electricity is sold, whereas the right to
walk on the pavement is free, is that there are meters to
measure electrical energy consumption but nothing equivalent to measure sidewalk use.

If redistributing wealth is a consequence of innovation,
then the likely direction of this redistribution is a key question to be asked of existing and proposed technology policies. In the absence of a specific commitment to promoting
equality, policies are as likely as not to move us in the
opposite direction – towards great inequality. In fact, I
know of no evidence that present government or corporate
technological policies are attuned to egalitarian values (even
though other policies, not directly related to technology,
and quite probably less influential, quite visibly are
influenced by the value of human equality). Thus, it should
be no surprise that current technological trends, to be
explored in the next chapter, are helping to polarize wealth
and even to undermine the myth of universal middle-class
equality that seemed to be guiding American culture and
politics from World War II until the last decade. A decisively different framework for technology policy will be
needed to turn this around.

Even at best, innovation policies can only allow choices of
which technologies to make available everywhere. Almost
inevitably, there will be some people who would suffer as
a result of even the best such policy, unless it includes
devising adequate compensatory measures before the
changes take place. Localities and individuals who lose out
because of innovations such as those just described rarely
would have been able to prevent or even foresee these
losses, and at the personal level, these changes can be devastating: the loss of a saleable skill mastered over a period of
many years is very nearly as tragic if it comes about as a
result of a new invention as if it is as a result of assault,
accident, or medical malpractice. While juries often award

adequate recompense on the basis of lost earning power to those hurt by the latter agencies, our society is curiously sanguine regarding the fate of those who suffer because of perfectly purposeful innovation. As the pace of innovation accelerates, justice in this area will be even more necessary. The crude solutions of retraining, temporary compensation, or early retirement that sometimes are offered ought to be replaced by much more compassionate and well wrought approaches that take into account just what the injury actually entails in each case.

Technology and power

The relationship between technology and power is complex. It will help to distinguish several different types of power. The first is simply human power relative to the natural environment, control over which is also control over the human future. Augmenting this type of power is both the most obvious – indeed, the most celebrated – value of technology and its most questionable consequence from an environmentalist point of view.

A second kind of power is mental power – the ability to perceive, remember, and draw conclusions – and this computers, television cameras, microphones, typesetting, etc., each augment to some degree.

The third, most important type of power from the social perpsective is – roughly – the ability to get what one wants from other human beings or to control their lives. This is what politics is all about. Since no one lives in a social vacuum, it would be impossible for each person to have full control over her or his own circumstances. Still, it ought to be possible to distribute power approximately equally. What equality of power would mean depends on the circumstances and so continually has to be redefined or – rather – renegotiated. When a political system fails to allow a renegotiation of equality, then it inevitably tends towards unequal power. Inequality is not just an abstract evil. As people lose control over their own lives, they lose the capacity to develop the kind of independent understanding of the world that can only be based on trial and error. Even when they do understand a situation, they are not in a position to act on that understanding. Those who have disproportionate power bear an increased responsibility for the suffering that inevitably ensues from this loss in effective social intelligence. Even if the sheer love of power hasn't already corrupted the powerful, the increase in responsibility always seems to justify both attempts to gain further power and fears that inevitable suffering will lead the powerless to seek revenge. As the distance between powerful and powerless grows, recognition of their common humanity becomes more difficult. Recent history is full of examples of the incredible gruesomeness that can result.

This third kind of power – over others – is infinite in its variety and gradations, from wheedling to using deadly force, from convincing through rational argument to misleading, and from bamboozling to enslaving. Technology enters here at times in extending the powers of the already strong, at times in boosting the powers of the relatively weak, and in adding to the horror when inequality becomes too great. Weapons systems as complex as aircraft carriers or nuclear missiles obviously are intended to strengthen the most powerful countries. Centralized, hierarchical systems of information channeling and communications are useful for augmenting the power of small groups of managers in multinational corporations or governmental bureaucracies. On the other hand, bicycles are likely to augment the power of large groups to intercommunicate, or to organize to resist centralized authorities. The Colt 45 pistol was known, with some truth, as "the great equalizer," because it was easy to obtain and a person did not need much physical strength to use one effectively.

Technological progress itself becomes a tool of power, as in an arms race or through the actions of a company like IBM, as it threatens to leave competitors in the dust. Continually new inventions always mean that the innovator has an edge on those waiting for the innovation to be distributed through the world. Thus, the European countries could afford to sell weapons to the inhabitants of Asia, Africa, and the Americas in the sixteenth through nineteenth centuries without fear of losing the upper hand because the Europeans would always have still newer weapons, better organization, and much more information. This effort still continues. The US and other industrial states hope to maintain their power relative to the Third World states despite selling them sophisticated weapons, partly through the one-sided capacity to develop still newer weapons the Third World countries will then have to obtain if they want to maintain equality – although these hopes may prove mistaken. Likewise, the US arms race with the Soviet Union has at least one stated purpose: to force the Soviet economy into such bad shape that it is no longer a match for the US. Similarly, continued industrial progress in general has the purpose of maintaining power over Third World and even advanced industrial countries.

New communications and computer technologies have been used to augment the power of relatively well-off social groups. Thus, televised "teleconferences" are much more available to the wealthy than to anyone else, and help them organize for power against others.

In more specific settings, such as workplaces, technology may also be used to augment the power of various groups. When computers were first being introduced into businesses, programmers were able to use their specialized knowledge coercively to raise their wages and control the pace of their own work, much as older skilled groups such as machinists have used their skills as bargaining tools. Management and owners have had two sorts of responses. First, they have tried to alter, or at least to understand, the technology in such a way that they would have more ability to control its use. Also, they have helped sponsor training programs and schools that widened the pool of prospective skilled workers, allowing owners and managers to select the more docile among them. In both cases, technology, in the sense of the knowledge of the practices involved in a particular endeavor, was crucial to the struggle.

The rapid introduction of new technologies also requires a corresponding upsurge in regulatory activities if the social and environmental fabrics are to survive. Often concealed within a seemingly neutral regulatory process can be decisions of far-ranging impact, therefore according substantial power to those in a position to affect these regulations.

Because technology changes what is possible on such a wide scale, control of technological decisions is in itself a substantial source of power, shared collectively by technologists and their employers.

A prohuman technology policy would seek technological paths that help equalize power rather than centralizing it. This involves not only a choice of technologies, but care in the patterns of distributing the new technologies. Both the resulting regulatory process and the direction of technology itself must be subject to democratic involvement, if true equality of power is to survive. For all this to be possible will probably require new democratic institutions which in turn could only work if suitable new technologies are devised; this development obviously deserves high priority, and great care.

Technology and culture

The word "culture" has many meanings, from a knowledge of the fine arts to the broadest possible way of describing a people's daily lives and patterns of thought. It is that large meaning that is now involved so intimately with technology. "Culture" in this sense describes our habitual patterns of action, both conscious and unconscious, our ways of proceeding towards goals, and the ways we conceptualize all these activities. Humans' tremendous repertoire of possible actions achieve pattern and meaning only within the framework of specific cultures. Loss of a coherent culture or lack of any role within one's culture can be as life-threatening as starvation itself – and is certainly as destructive of humanity. Less extremely, a culture can become too thin, so that the inventory of acts with shared meanings shrinks within it, while personal and community ties fray towards unrecognizability. This can happen when people from diverse peasant cultures are forced into rapidly expanding urban slums – a common pattern today in the Third World.

On the other hand, pressures to obey traditional cultural patterns are also often debilitating, as they require narrow restrictions of personal choices and limit capacities to deal with circumstances that inevitably continue to change. Often the call for supporting traditional culture serves as a guise by which to justify and perpetuate sharp inequalities of power, wealth, or opportunity.

There are at least four areas in which technology influences culture directly and profoundly. First, in the realm of what, in modern society, we call work or labor – our culture is totally tied to the technologies that shape the work environment. Recall that technology is more than tools and machinery; it encompasses the engineered and standardized practices for operating that machinery to some desired ends. If culture is the way we organize and understand our actions, then in many work settings technology very nearly is culture. In addition to the practices established by engineers and managers, most work settings have a recognizable "informal culture" of interaction among workers. Even the informal culture is strongly dependent on what technology dictates: "formal" interactions of cooperation, competition, or hierarchy among workers, distribution of tasks, physical activities, range of skills needed, how much and what kinds of prior training, experience, and learning are required for each, spatial layouts, work rhythms, the degree of routine, the chance of escape or rest from a task, intensity of concentration required, workplace location, size of a workplace, the relationship between workers and customers or clients, levels of noise and fumes, and at least in part even relative wage structures for different kinds of tasks, relative status of different skills, and job security.

(Some of these aspects are discussed in more detail in Chapter 8.)

Whether in a factory or a bank, if a task requires close cooperation and a high degree of attention, it will fit with certain styles of camaraderie much better than with others. If it requires no cooperation but high attention, it will isolate workers even sitting right next to one another. These patterns quickly become habitual, and they can alter ways of experiencing one's fellow workers, and therefore influence ideas about other people in general.

A second range of influences of technology is personal life, the life lived at home, among friends and family. A well-known example of the role of technology is the way marriage patterns changed in the French countryside after the advent of the bicycle in the late nineteenth century. Young peasants now had a means to travel ten or more miles and back within a single day, while still having the energy to work or to court. This greatly broadened the number of potential available spouses; rather than coming from the same or the next village, spouses now tended to come from considerably further off. As a result, the family into which one was marrying was far less likely to be known to one's own family. The whole set of criteria upon which choosing a husband or wife was based could change as well. What was true for bicycles is even more the case for cars, and now airplanes, and what was true for courting couples is also true for friendships and acquaintanceships. Technology has had profound impact on patterns of intimacy, on the possibility of choosing one's associates, and as a partial result, on the difficulties attendant upon the power to choose.

In addition to the forming of personal ties, technology influences home life through all the appliances, entertainment devices, furniture, and other products that have come to be taken as necessities in any well-equipped home. Familiarity with the operations of all of these and facility in using them are necessary signs of acculturation in our society. Proper upbringing is demonstrated by the "correct" manner of squeezing the toothpaste tube, replacing the phone on the handset, putting water in the ice tray, or of tuning a television, and who thinks of any of these as imposed by technology or – still less likely – by the politics behind technology?

The home, too, generally still taken to be women's sphere, is replete with labor-saving devices that have, we are told, taken the drudgery out of housework. In one sense that is true: gruelling tasks such as doing laundry entirely by hand are gone, but with new ease of performance have come additional assumptions about what constitutes good housekeeping. Each new item brought into the house comes with instructions about how it is to be used, and if not to be immediately consumed, how it is properly to be stored, cleaned, repaired, and maintained. These instructions codify the requirements imposed by technology. To the extent they are followed, the home can become a scene of direct control by technologists to almost the same degree as the factory. We have arrived at the point where gender and technology interact.

Gender relations and technology

In virtually every known culture, tasks and social roles are divided along gender lines (although, it is important to remember, in no two cultures are the lines the same). There may be cultures for which it is possible to argue that such a division does not (or did not) involve the oppression and exploitation of women. But that is certainly not the case in our own society. Here, the division of roles according to gender, often tied to particular technologies, does tend to leave women in lower-status positions, lower-paying jobs, and, increasingly, raising children alone and in poverty.

Just as in traditional societies, certain tools are taken to be women's, whereas others are men's, so in our own system even recently developed tools somehow seem to have gender attached to them. Thus, the large-scale introduction of typewriters into offices coincided with the introduction of women into clerical roles. Similarly, the position of telephone operator or receptionist early became identified as female. In the case of the telephone, there appears to be an accidental technological reason: the original system was particularly suited to the clear transmission of sounds in the range of normal female voices. To a lesser extent, that still holds. However, electrical engineering long ago reached a level such that, had a conscious effort been made, this peculiar gender restriction could have been overcome. That that is not what happened indicates satisfaction – on the part of those who controlled technology – with the existing division of labor.

As everyone is aware, among the gender divisions of occupations, technology as an activity is almost purely male. It is little wonder then that there has not been much consciousness of possible contributions to sexism in technological choices. (One area where damage resulting from this has been most severe is in international development programs. There, often even otherwise very well-intentioned efforts are marred because decisions affecting what has been traditionally women's sphere are made without consulting them. For instance, gathering firewood is often a female task. Developers consulting only the men in a village might conclude they would benefit from replacing nearby forests with orchards producing marketable fruit; but since the men do not necessarily think of women's role at all, the actual result could be a serious shortage of firewood and much harder work for the village women.)

The lessons for a technology policy are that greater efforts to develop technologies that do not carry hidden and harmful gender implications are needed, that ways must be found to involve women more fully in technological decisions, and that strong efforts to change gender balances among technologists are of great importance.

Ethnic relations and technology

Much of what has just been said about gender applies to ethnic divisions. In many traditional societies members of minority or foreign cultures often were limited to or had a monopoly on certain occupations. That practice has carried over to a considerable extent in our own society. At times, "labor-saving" technologies can single out for elimination those already "undesirable" jobs, such as farm labor, that now are reserved for minorities. In addition, because many technological choices are made with the majority culture in mind, members of ethnic minorities face an unnecessary handicap in having to adjust to some technologies. And, of course, other "minorities," such as the left-handed or the color blind, are often ignored in a similar way.

The public sphere

Technology affects the portion of culture we might call "public life." In a way, the very existence of an extensive public life, separated from personal life, is an aspect of modernity still new to most of the world. Such activities as attending school, shopping, traveling, vacationing, movie-going, and voting are part of this sphere. In it, old bonds of family or familiarity are loosened, and ties tend to be casual, limited, or at times virtually nonexistent – as in the degree of rapport one might feel with a driver of another car on the same freeway or with a bureaucrat one expects never to see again. Technological changes often facilitate and even propel the social changes.

Moves towards anonymity are by no means all negative. As it originated and grew, the public sphere opened up new possibilities for cultural richness and diversity. It came to be the scene for innovation, expression, and new forms of discourse. The relatively powerless, meeting together in the partial anonymity of the city, were able to form new organizations and unite to press their demands. More recently, standardized, computerized means of voter registration have helped prevent the sublegal practice of systematic exclusion that confronted Southern blacks before the 1965 Voting Rights Act.

But the growth of anonymity can cut two ways. Centralized retailing with automated cash registers has greatly reduced the power of the local small business or the independent-minded clerk to extend personal credit or even outright gifts of merchandise to neighbors in distress. In the modern city with automated banks and mass transit ticket machines, the need for personal contact may be reduced to changing the ten-dollar bills from the banks to the five-dollar bills accepted by subway systems, but that transaction can become desperately difficult; it has become too big a favor, too big a risk in a world of strangers. Cheap pistols are one of the technological hazards of urban life, but they are only one reason for suspicion and guardedness that alter contacts in public. As the use of bank cards, credit cards, etc. to establish identity amid anonymity grows, so do the possibilities of counterfeit, fraud, and theft that place us further on guard.

As was suggested above, in the fairly recent past there were public spaces – taverns, churches, community parks, coffee houses – in which even the powerless could meet. The union movement, for example, was strengthened by the existence of communities in which workers in the same industry, but with different employers, worked, lived, and met together. Subsequent changes, among them the rise of the automobile and electrification, played a major part in permitting the reshaping of the social landscape that has removed most such public spaces and has dispersed communities. Reduced possibilities for meeting people with common political and cultural interests, and a lower overall level of human interaction, is a continuing weakness of contemporary public life. Helping assure a capacity for participation in genuine communities of some sort ought to be one of the important goals of a worthwhile technology policy.

Technology and geographic relations

Technological possibilities, combined with other social forces, alter numerous aspects of geography in addition to the urban landscape. For those who can afford it, it is now possible to telephone thousands of miles as easily as fifty feet. The main time delay in an intercontinental phone call may well be the time it takes the person you are calling to pick up the phone. Our concepts of distance have been strangely inverted. Distances in a room loom larger than distances between countries. Through frequent telephoning, electronic mail, and air travel, managers or professionals within an industry but working thousands of miles apart may be closer than the ordinary workers in that same industry – even within the same plant – are to each other. Social and class alliances, identification, and cultural norms all change as new technologies redefine proximity.

The inversion of geography does not end there. We are familiar with the fact that travel across a metropolitan area may take nearly as long as a coast-to-coast flight. As some travel becomes easier, other distances get relatively longer. In many parts of the Third World, the distance from city to country has hardly shrunk at all, while the well-off in the cities continue to move closer to Europe, the US, or Japan. Existing political barriers to travel and communication increase relative distances between countries even as technical barriers have fallen. This leads to especially disturbing consequences in relations between the US and the Soviet Union.

Both the US and the USSR have experienced themselves shrinking as countries, as internal travel times continue to drop and as the rest of the world has grown in power. Simultaneously, the travel time for deadly weapons that could be launched by one against the other has gone down from hours to only a few minutes, and shows every sign of continuing to decrease. But the visa barriers and other impediments to free travel and even to telecommunications between the two countries have hardly changed or have gotten worse. In effect, each country is seeing its own power shrink and feeling increasingly threatened by the other, while, at the same time in terms of personal familiarity, the two countries are flying apart. Western Europe, too, is now closer to the US and effectively further away from Russia than has been the case for centuries. These perceptual changes amount to actual changes in relations that are obviously extremely dangerous. Yet, the problem is less that political barriers have actually gotten worse than that they have not diminished as fast as other barriers to travel and communications.

Technology and "comparative advantage"

Technology has a general drift of denying the particularity of place, of group, or of person. Whereas in premodern society, each place had its own way of doing things, each occupation its own skills, etc., our society is moving towards a situation in which everything is completely interchangeable. Damascus steel, for example, was once the secret of India. No one but Indian steelmakers could produce the metal. Now the technology of steelmaking is universal. A steel plant can be built wherever it is wanted. Inhabitants of the area can be trained to carry out the carefully devised tasks of the modern mill. The same holds for automobiles, for computers, for almost any sort of food product, for almost all activities.

Where a region seems to lack specific resources or a favorable climate for a certain product, technology offers alternatives such as substitute materials or processes, means to alter climatic consequences, ranging from vaccines to irrigation to air conditioning; and, if substitution is not possible, relatively cheap transportation of low-cost raw materials from a variety of international sources. To have a steel industry, a country need not have its own iron or coal resources, to have an automobile industry it need not have a steel industry, etc.

When there is a craft or skill that cannot easily be learned, the resulting product or service may often be mimicked by one resulting from a more technologically-based process. For instance, authentic Persian carpets are each individual products of particular Iranian tribes. They are hand-knotted according to designs, the details of which are made up as the knotting proceeds. Each rug has visible irregularities of pattern symmetry, of design, of color, of shape. All this complexity can be reproduced with ever-greater fidelity by modern rug-making factories in which the patterns may be printed or woven, or possibly even machine-knotted. The basic rug-making patterns can be analyzed with the aid of computers; variations and even irregularities can be programmed. It becomes increasingly difficult for anyone but an expert to distinguish the imitations from the real thing. As objects for ordinary use, the imitations are just about as good. But the imitation rugs can come from wherever there is a rug factory with modern machinery.

This tendency towards being able to do anything anywhere, which is by intention an inherent trend in technology, has a strange corollary: there is less and less reason for trade as we understand it. What makes this paradoxical is that technology has simultaneously facilitated an ever-larger volume of trade. In fact, the economic profession's overwhelming consensus is that high levels of international trade are greatly to be desired.

The basis of support for high trade levels is the theory of "comparative advantage." In essence, this theory states that if an area concentrates on producing whatever it can produce most efficiently, and trades those goods for others it could produce less efficiently, then everyone will be best off, because more goods will be made available than would be possible any other way. The theory has several blind spots, but even before considering them, we can see that the trend of technology is to eliminate comparative advantage. There is no reason why one region should be very different in its relative efficiencies in producing different things than another region, since both could, in principle, use comparable technology for producing each thing. The real difference in relative efficiencies can only come about if one region is not allowed to have the same technology as another. Aside from that, greater efficiency really only means being able to produce the goods at a lower price.

With comparable technology, what being able to produce at a lower price means is lower wages. The rise in international trade really ends up forcing workers' wages to be set at the lowest possible level. Since workers in a country like the US have only been able to achieve high wages through long labor struggles, political battles, organizing efforts, and so forth, international trade is to a considerable extent a means of undermining those hard-won victories. Korean television assemblers are not strikingly more efficient at producing television sets than American workers with comparable equipment. Their real efficiency is lower wages. Korean workers' wages can be low partly because even with these wages the workers can buy more goods than they are used to having. But a large part of what keeps them from wanting more is a repressive regime in which serious unionizing or striking for higher wages would be very difficult.

High levels of international trade, then, depend on keeping workers' wages low, but keeping wages low means demand for goods is lower than it would be otherwise, so low wages mean less employment as well. (Likewise, high levels of unemployment mean strikes are dangerous – the strikers can easily be replaced – so wages stay low.) Wonderful as international trade may seem for those whose livelihood comes from exported goods, or from the activities of trade itself, the value of the whole system is increasingly questionable. A growing minority of investment studies reveals that, with certain exceptions, a sounder plan for economic vitality would be for communities to try to meet their own needs. The minimum advantage gained is saving those resources currently needed for trade and transportation.

Within a community of moderate size, economic resources can be adapted to existing needs and desires faster and more efficiently than when the entire globe is involved. With economic self-sufficiency, a community is also in a much improved position to gain political independence in a meaningful way as well; choosing to devote resources to adequate welfare levels, environmental improvement, or cultural survival would be much easier if it were not necessary to be competing with areas that chose not to do this. The trend of technology would certainly suggest that this goal of economic self-sufficiency is increasingly feasible. (Such communities need not be geographically contiguous areas; modern communications permit many different ways to organize self-defined communities.)

The exceptions are important. The major one is that to make it possible for each region to live at the level it would want, there must be a "transfer of technology." Since "technology" has several different meanings, it is not at all surprising that the concept of technology transfer is a fuzzy one. In one interpretation, sending a car to Uganda could be considered technology transfer. At another level, the phrase could describe shipping an automobile assembly plant to Brazil. At still another level it could mean sharing the very latest technology of the design of automobile plants with Japan. Finally it could refer to helping another people solve a production problem in its own terms, and it is this last sense that deserves more emphasis. (For example, certain tribal agricultural societies are able to live at fairly high densities in tropical rainforests by using a method of agriculture at the opposite extreme from what we call "scientific agriculture." High efficiency in American agriculture involves single crops with extensive use of pesticides, herbicides, fertilizers, and complex, motor-driven machinery. The tribal agriculturalists, by contrast, work in small clearings where they grow up to eighty crops together in an intricate pattern that produces substantial yields while minimizing the problem of pests or the need for fertilizer. Yet our biology could be used to analyze their basic methods and to suggest how, with a minimum of extra work, these native farmers could further improve yields or prevent slow decay of their environments. A very tiny percentage of Western agricultural scientists have helped in this manner, instead of offering the usual choice of total "scientific" agriculture or nothing.)

A viable international economic system would include community-based economies and free exchange of necessary technological information, along with intercommunity aid in emergencies. The size of each community would depend on the scale needed for fairly efficient production; as we shall see in the next chapter, that scale is now shrinking.

Technology and times of transition

Hidden in much of the current wave of interest in high technology and in the idea that "sunrise" industries promise economic salvation is a notion that past economic crises and subsequent recoveries also corresponded to the emergence of new technologies. The origins of this notion appear to be the suggestion that so-called Kondratieff long waves (cycles of economic boom and bust lasting for about forty years) are correlated with fairly major technological shifts – sometimes referred to as the second, third, and fourth industrial revolutions. Since there is some evidence supporting that hypothesis, it is necessary to try to understand the phenomenon a bit more deeply in order to gauge the extent to which it might apply to the present and near future.

In the framework of a particular means of distributing buying power (e.g., factory wages, welfare), a definite set of technologies permits the elaboration of a certain way of life – for a time. For example, in late nineteenth-century America, life had come to center on the railroads that brought grain and meat from Midwestern farms, and coal and ores from more scattered mines, to industries in midwestern and Northern Atlantic coast cities. Then, as farm and factory production continued to expand, incomes available to buy the products expanded less rapidly. Rising productivity per worker and farmer meant that fewer farmers and workers were needed to produce the currently desired array of consumer goods. The remaining farmers and workers were then less favorably situated to press for higher incomes. Meanwhile, monopolistic industries had little motivation to lower their prices. As consumer demand stopped rising, further investment was not warranted. Businesses producing goods for other businesses and for farms began to fail. Workers in these industries lost the buying power they had had, and the whole system began to collapse. Uncollectable loans led to bank failures and finally to the panic of 1893, which was followed by prolonged depression.

Several factors helped bring about at least a partial recovery. New technologies such as electricity, the telephone, and, a little later, the automobile began to permit a major reorganization of life. Demand for these new types of goods was not as limited, since the wealthy and those who still had jobs could switch their allegiances to the new products. The industries associated with the new products could expand, and therefore they became a market for new capital goods (including copper wire, telephone poles, etc.). The new technologies also permitted a rearranged urban landscape: industries operating with electric motors and lighting could use factories different in shape and location from directly steam-powered, sunlit ones. Electric trolleys and then automobiles permitted the middle classes to move to nearby suburbs. National consolidation of industries was aided by the telephone. The alterations in urban forms involved increased demand for construction workers and for the materials they used. Accompanying these physical changes and helping make them possible were a variety of governmental changes which are often characterized by the name "Progressive Reforms." They included reform of urban "machine" governments, new, large-scale public works projects, public health measures, and new sources of finance and capital. The sum of all these changes led to renewed economic growth, growing business confidence, and a new rise in employment. (In the midst of this process, and related to it, came World War I, which for a while further stimulated growth.)

What exactly was the role of the new technologies in all this? Did the crisis lead to an upsurge in invention, or were the new technologies just sitting and waiting for a chance in the market? Neither answer would be correct. The chronology of inventions themselves does not have a particularly sharp peak corresponding to the post-1893 recovery. But the older industries were "mature," in the sense that without a major restructuring of the world distribution of wealth, there simply was little likelihood they could grow further. Because productivity and efficiencies of scale were already high in these industries, small growth of demand for them was not going to result in overall economic growth. In part, that was because the world had already been shaped to accept the output of the existing industries.

Jobs that had initially been necessary in establishing the railroad-era way of life – for instance, in clearing areas for farms near railroad tracks, or hauling wood by cart to build tracks – were no longer important or very necessary. But the new technologies which had recently been developed – electricity, telephones, etc. – could be utilized effectively only if people changed their way of life to accommodate them. This process took time; while the changes were under way, a period when both new and old technologies were required would mean high levels of employment. Furthermore, the new technologies created new sources of wealth – for instance, farmland that suddenly gained in value because of its accessibility to automobile traffic. The owners of this newly-created wealth were able to convert at least some of it into buying power, thereby increasing overall demand levels. At first, too, the new industries – both those delivering the new technologies and those facilitated by them – were highly competitive. They operated without substantial profits, but with relatively high employment levels. As the new industries came to displace those of the prior era more fully, and as they once again became highly concentrated – each one dominated by a very few firms –

the conditions that would eventually lead to another crisis of overproduction and underconsumption were reestablished.

Technological change, according to this summary, can lead to economic rebirth only if it is part of a larger "structural transformation," including cultural, political, geographic, and financial changes. The concept of an overall "structure" suggests that we cannot change just one important aspect of our lives, while leaving other aspects intact; changing one part of a structure requires changes in other parts as well, if the whole is to continue to hang together. New technologies can have value only if we want the other changes that would go along with adopting them.

If we decide to limit new technologies to those that could be adopted without other significant changes in our way of life, then, in effect, we could only be replacing some industries with others very much like them. There would be little point to such a switch, since there would be no reason for it to make much of an economic difference either.

This argument does not mean that there is necessarily only one viable societal structure consistent with a given set of technologies; there may be many, or there may not be any. When enthusiasts of high technology see it as saving our society, they are implicitly postulating structural changes, but are silent on just what the extent and direction of such changes might be. We should not take it for granted either that these changes will, or that they will not, automatically lead to a better society from the standpoint of humane values. It is equally unclear whether there can be a viable new structure involving high technology that would preserve intact such central features of our society as the market, private property, personal identity, and other aspects of social structure which prior "industrial revolutions" supposedly left more or less unchanged.

The next chapter sketches, in brief outline, just what kinds of structural changes adopting high technology might entail. That will allow at least beginning to frame the political question of which changes would be worthwhile.

4 High technology and its possible impacts

What is high tech?

The focus of current political interest is high technology – "high tech" for short – which is the characteristic technology of our time. Exactly how it differs from ordinary technology is open to debate. Standard definitions often are purely numerical: a company is considered high tech if the percentage of engineers among its workforce or the fraction of revenues spent on research and development is well above normal. Such calculations fail to capture what makes high tech new and different; for instance, they would include companies that design quite standard dams or bridges. Even the interior design aesthetic of the same name suggests there is more to it than that.

It would not be appropriate here to attempt a complete definition of high technology, since that could only be accomplished by detailed exposition of numerous examples. Yet, to understand high tech's potential social impacts does require some grasp of its fundamental attributes. High technology is the outcome of the enormous expansion of science and technology in this century, and of their increasing interpenetration. In the period after World War II especially, that expansion was fueled by national governments and – to a lesser extent – by private corporations sponsoring research and development on a vast scale in pursuit of four basic goals. In descending order of importance these four goals were: military power, prestige (as in the race to the moon), commercial competitive advantage, and cures of disease. Those efforts built upon all past science and technology, but most especially upon the unified picture of the structure of matter and radiation at the atomic level that emerged in the first half of this century.

While nineteenth- and early twentieth-century observers were keenly aware of the importance of science and technology, the growth of these fields has been so fast that in some ways it is accurate to say that the scientific and technological eras have just dawned. The exponential growth of the number of scientists is such that the majority who ever lived are still alive. But beyond sheer numbers there is much more: the international scientific and technological communities are connected as never before. On the whole, the cross-fertilization among disciplines and subdisciplines occurs at a faster pace than ever. Individual scientists and laboratories are in general more productive than ever because of the variety of both instruments and cognitive tools available to them. Finally, research has been planned, parceled out and administered with ever greater sophistication. Despite the fact that details of such planning have definite drawbacks, they have led to very rapid movement into new areas of investigation, and to equally rapid conversion of results into technologies of value to the institutions sponsoring the research.

Among the reasons for this recent rapid growth have been certain key technologies, especially digital computers and the associated computer programs. They began to emerge immediately after World War II and quickly became central to the functioning of military and then other laboratories. Additional technologies such as telecommunications and jet travel to international conferences have helped tie international scientific and technological communities together as never before. By now, university education of foreigners in science and technology is a substantial item of international trade, even if it sometimes is not counted as such. One of its effects is to spread throughout the world common habits of laboratory practices, and also of perception and conception so that both science and technology have reached new levels of international uniformity.

If we accept that science is largely the gathering of data, and the separation of interesting data (from the experimenter's viewpoint) from everything else, then more sensitive instruments and computers have made possible million- to billionfold improvements in many fields, just since World War II. This says nothing about the diversity of the information or the sophistication of related theories which have emerged as a result.

Despite the difficulties of description, it is possible to propose a few generalizations relevant to the character of the high technology that has emerged.

1 In many areas, science and technology now affect each other so rapidly that the borders between them have virtually disappeared.

2 The greater understanding of matter – including living things – at the microscopic level has continued to grow in sophistication, and now permits a technology involving extremely detailed reshaping of matter on the same scale.

3 It is that microscopic reshaping that more than anything else characterizes high technology. There are three chief areas where it plays a role.

a *Technologies of information and communications*, exemplified by microelectronics, optical fibers, and satellite communications. Information technology has burgeoned enormously. Along with the hardware, there has been a parallel and perhaps even more substantial growth in the "software" – computer programs, organizing principles, images, and data. (In fact, it is arguable that the sum total of computer programs in itself constitutes the single greatest outpouring of new technology in history.)

b *New materials*, including plastics and other organics, metals, composites and ceramics, and prepared surfaces, all designed with a growing sophistication on the atomic and molecular scale. These materials are in most cases only at best slightly superior in their basic physical properties (strength, melting point, magnetizability, etc.) to older materials, but they often are cheaper and easier either to make or to shape into useful forms. They also can be made remarkably uniform.

c *Biotechnologies*, including molecular medicine, genetic engineering, new cloning methods such as hybridomas, new methods of investigation such as fluoroimmunoassay or nuclear magnetic resonance computer-assisted tomography.

In addition to the three leading areas of high tech, there are many others, probably of less social importance: space, ionic plasmas, ultrasonics, high-power lasers, maglev transportation, breeder reactors, etc. In all these areas, there is often a requirement of laboratory-like precision and control over conditions. Further, most of these high technology areas involve the interplay or integration of several heretofore separate sciences and technologies.

High technology involves remarkable scientific and technical achievements. It first appeared in laboratories, in space, and in military systems. Only in the last decade or so has it begun to have much direct impact in our daily lives, and that impact is growing. Its political relevance depends on the degree to which it can permit a reshaping of our lives. There are two aspects to this question. First, given current trends, what kinds of reshapings do we seem headed for? Second, if it were possible to redirect technology according to more humane values than currently dominate, what would these new technologies permit that prior technologies wouldn't have?

To address these questions, it is useful to start with a brief overview of the nature of possible changes, and then to turn to an examination of possible impacts roughly according to the categories set out in the last chapter: wealth, power, culture, gender, geography.

The chapter concludes with a brief and tentative answer to the question, posed at the end of Chapter 3, of whether the new technologies and other current trends seem to be leading towards the emergence of a new, coherent, economically and socially viable "way of life."

An information-centered society

All three of the major new technologies help propel us towards an information-centered society in which most people would identify their most important activities in terms of obtaining, working with, or expressing information, whether in the form of entertainment, data, instructions, computer programs, records (including medical records), or various forms of telecommunication. It is not surprising that this should be so. Information, as it is not material, has no necessary size; high technology is most directly related to the manipulation of matter on a scale smaller than that on which material objects of direct use to us are built (i.e., housing, means of transportation, furniture, clothing, and food all have to be of a size related to human size.) Accordingly, many of the most important new materials are primarily of use for microelectronic and related devices such as liquid crystal displays. Biotechnology includes many tests and assays that primarily increase medical or biological information. Even such biotechnologies as genetic engineering conform to the paradigm of information-related activities: genetic engineering involves obtaining the gene-coding for particular molecules within a living thing; the code may then be transmitted as information to a distant laboratory which can then re-encode it in DNA and insert it in a cell to start a biological process.

Changes in production

Even in an information-centered society, material goods would of course still be needed. But the nature of production could well change in several important respects.

Information technology lends itself to the automation of many tasks, and to the elimination of other work through using the coordinating power of computers and telecommunication. High-technology items themselves often require so much precision that only automated equipment can produce the necessary tolerances. Moreover, many complex electrical or mechanical components, such as gear trains, the manufacture of which in the past required intricate machining and assembly work, can now be replaced with much more standardized microelectronic controls, along with easily changed programs; as a result, manual production is much simplified. The new materials, and even biotechnology, also permit eliminating many skilled steps needed in prior production practices; for example, certain new metals or plastic or ceramic replacements can be shaped through molding, casting, or forging to such precision that little skilled machining is then needed. Biotechnology can replace highly complex and exacting chemical processes with far simpler "brewing" of bacteria or yeasts in relatively simple bioreactors.

Because telecommunications between computerized systems permit coordination of operations at different points, production processes can be decentralized. Multinational corporations now are routinely able to divide production tasks for certain products among many different countries, depending on the degree of skill and the cost of labor and other resources associated with each stage. Usually, these separate tasks may even be shifted rapidly from country to country. At the same time, direct links with the ultimate purchaser now allow dispensing with many intermediate stages, such as the warehousing of goods at the regional, wholesale, and retail levels. This lowers total production requirements, and reduces needs for resources as well.

Automation and decentralization, or the transfer of work to remote places, is not limited to material production tasks. Service work of all kinds, from banking to health care, is subject to many of the same sort of changes.

Without a shift in dominant values, pressure for automating and decentralizing of production and services is hardly likely to let up. One reason is that the business of devising new technologies for this purpose is constantly

attracting new entrants ranging in size from individual entrepreneurs to the largest corporations. This, combined with
rapid progress typical of fields such as computer programming and microelectronics, strongly suggests that the pace
and scope of transformation of workplaces and jobs are
likely to continue to increase rapidly. Without accepting
claims that computers will soon reach human-like intelligence, it is perfectly sensible to believe that, by as soon as
the end of the century, if these trends continue unchecked,
the number of jobs seriously affected could grow from
perhaps tens of millions in the world today to billions – that
is, virtually all current, repetitive jobs.

(The number of jobs likely to be affected would not
necessarily equal the net rise in unemployment, not only
because some good new kinds of work might emerge, but
even more because, faced with dire poverty, many people
will be willing to take on whatever work is open to them,
no matter how demeaning it is, or how low the pay. See
also, below, the section entitled "Work culture.")

High technology's potential for redistributing wealth

High technology can affect the distribution of wealth in four
major ways. First, the single most important new category
of wealth it potentiates is information in its many forms.
Second, new materials and biotechnologies, along with
information technologies, undercut the value of many
existing sources of natural resources, since they allow the
replacement of one material with another, or permit the
more efficient use of already available objects. Third, as we
have already seen, the potential of labor-saving technology
reduces the value of many skills. Finally, the enormous
research effort needed for high technology makes research
capabilities themselves a potentially important form of
wealth. If current trends continue, these shifts all would lead
towards further concentration of wealth in the advanced
countries, since raw materials have been the principal source

of wealth for the less developed countries, and since the advanced countries alone have the means to collate and comprehend the new wealth of information.

Within the advanced countries, also, current trends suggest that, despite new ways to make information widely available (e.g., personal computer networks, larger computer banks, interactive video, and computerized voice-data communications), it is likely to be so only in a useful way for professionals, managers and a few other white-collar workers, largely for the benefit of corporations and governments. Research institutions, whether private or university-based, will be sources of wealth to those who control them. Meanwhile, skills of many manual and white-collar service workers can be expected to decline in value. In other words, there is a large segment of the population from whom wealth is being redistributed away.

Property

As was discussed in the previous chapter, wealth is converted to property as the consequences of social convention. Currently, information in the forms of both expression and technological knowledge can be held as property, as a result of the intellectual property laws (to be discussed in more detail in Chapters 5 and 10). Since new information technology includes easy ways of reproducing information, the existence of these laws effectively curtails the widest possible spread of this new form of wealth. Unlike material objects, information can be shared widely without running out. Therefore the intellectual-property laws help create a distribution of wealth that in some sense is unnecessarily limited. Furthermore, since it is possible to steal intellectual property in the comfort of one's own home, or in the course of using the communications system, enforcement of such laws entails a degree of surveillance and infringement of privacy quite unlike property laws affecting material things.

An alternative handling of the property question could

lead in a substantially different direction: with social wealth chiefly comprising information, equality of wealth becomes more imaginable than ever before.

Money

Money has been the standard measure of wealth, as well as the main vehicle of exchange, throughout the modern period. In an information society, the role of money becomes problematic from at least three standpoints. First, there is no feasible way to set a fixed value on information. It cannot be measured and doled out in standardized amounts, as raw materials, manufactured products, or even services such as garbage collection can. Second, high technology permits a flexible form of manufacture and service, such that each item produced can be unique. Its price is then arbitrary. Third, as financial institutions decide to speed their procedures, they convert money itself into pure information – with funds transfer involving nothing other than coded signals. There is no absolute way to protect such signals from a variety of manipulations which would involve instantaneous and virtually unrecognizable changes in the effective amount of money in circulation; many of these ways are now considered to be perfectly legal, and they have been adopted by financial markets (e.g., the proliferating variety of classes of stocks, of stock futures, letters of credit, and "junk" bonds). The result is that money becomes an arbitrary measure in an ever-growing variety of circumstances. The further we go towards an information society, the more likely it is that the effect will be to concentrate wealth in the hands of those most skilled at the appropriate manipulation of symbols – a class different from those who have been able to concentrate wealth in their hands heretofore. To avoid this, nonmonetary methods for the equitable division of social wealth would be needed.

High technology and social power

"Knowledge is power." To the extent that cliché remains true, information technologies have much to do with the distribution of power as well as wealth in our society. Through their diversity, these technologies permit large organizations to have access to a far wider range of information than in the past, and to be able to use it in turn for further strengthening themselves – especially when their operations are highly decentralized geographically. The top management of a decentralized corporation can be flexible, for example, about its sources of supply. Should one supplier plant have a strike, it is easy for management to obtain equivalent supplies from other sources. Since the workers in the plant are unlikely to be equally well connected to other workers, the power of labor in this situation diminishes sharply relative to those situations in which, because of greater communications difficulties, management is less flexible. Emphasis on information in society empowers those who are versed in understanding and analyzing information and have the time to do so – only a minority of the population. To extend democracy – i.e., to equalize power – would require extending opportunities for the necessary understanding.

New information technologies such as cable television and home video recorders apparently lessen the power of the few corporations that control the major networks to decide the flow of issues and concerns reaching public attention. But what happens in practice is that the media divide into two categories – those that keep trying to reach essentially everyone (i.e., the mass media), and more specialized media focused on sub-audiences. The majority of the specialized media will be aimed at "upscale" audiences who can afford either to patronize them directly or to patronize the advertisers who subsidize them. The view of the world that they will be interested in projecting is one that accords with the outlook of the already well-off.

The mass media, on the other hand, are increasingly

integrated into a kind of homogeneity by their communication with one another. They come, by a form of consensus-building, to agree on what constitutes news or entertainment. Because they all now feel obliged to cast sidelong glances at one another, each making sure it has access to what the others have decided is the major news, they are more subject than before to manipulation by newsmaking sources, such as the White House. Diversity of media does not translate into a comparable diversity of viewpoints. For that to be true, there would have to be substantial effort to design and build communications systems that exclude no one.

The nature of high technology admits of rapid proliferation of new variations. For example, new, special-purpose computer programs are endlessly appearing. Current methods of distributing innovations assure that the wealthy have a permanent edge. As long as these programs can augment in any way the organizing capacities related to power, then power will be unequally distributed. The preservation of democracy must, now more than ever, include equalizing the distribution of new capabilities.

In the field of military power, high technology is now looked to for a special kind of edge. "Smart" weapons and the electronic battlefield, etc. are means to try to assure that the armies of advanced nations can fight without casualties, thus permitting them the possibility of victory against determined nationalists fighting to protect their own soil and feeling themselves sufficiently powerless and degraded that they are willing to die rather than surrender. However, the world-wide pattern of arms sales renders this automated-weapon strategy problematic. As the British discovered in the Falkland Islands, the Argentinians had already been able to buy a "smart" weapon that the British themselves had helped design. In a world where alliances and adversaries are not fixed forever, technologically-based weapons may end up providing little advantage. The answer implicit in the "Star Wars" (Strategic Defense Initiative) and similar programs of spending an ever larger portion of the national product on weapons development so as to outrace the competition is likely to be equally pointless as a means to increase power for the US. But the effort will likely result in the further entrenchment of a military-industrial complex and, through a heightened arms race, increased chances of war resulting from miscalculation.

Giant information technology-based weapons systems, such as the proposed "Star Wars" system or the current "command and control" system for nuclear war-fighting, have two novel consequences as a result of their reliance on computer programs of unprecedented complexity. In both cases, these programs would be far too intricate to be completely understandable even by their designers; there would simply be no way to make sure that all possible serious errors leading to war by mistake had been removed. (In this they are like any highly complex technology – as, e.g., a nuclear reactor – in which the complexity makes it simply impossible to consider in advance all possible combinations of circumstances which could lead to severe malfunctions. In the case of the military systems, the consequences of such errors could be unimaginably devastating.)

The other novel aspect is that the key elements in designing such systems involve computer programming; this may be classified as "research" as well as "development." Once the programming or the difficult problem-solving involved in it has been done, the step to deployment is in effect largely accomplished; the program has only to be copied and added to less important material components of the system. Hence, there is reason to be wary of current statements that research – but not building the "Star Wars" system – is all that has been initiated so far.

Cultural implications of high technology

If it is true that all technology as a way of prescribing human practices has direct cultural implications, it is true with greater intensity for high tech. The actions that form part of any culture do so because they have meaning within the culture – whatever the action is, it also specifies a whole set of relationships between the doer and the rest of that cultural world. Baking a cake, carrying out a task on an assembly line, walking down the street, or taking a chest X-ray are all material actions, but they are also symbolic actions within our culture. As we move towards a way of life in which the primary activities directly have to do with information – that is, with symbols – the dual nature of activity tends to collapse. Furthermore, it no longer makes sense for our actions simply to reiterate existing symbols. Activity becomes the creation of new information, and that, in effect, means new symbol systems – new culture. More than ever before, we are in the business of constantly and rapidly transforming the shared meanings that have allowed us to make sense of our lives. Clearly, it is exceptionally difficult to foresee what such unprecedented fluidity and uncertainty would mean, so the following remarks are a highly tentative listing of some areas where major changes seem likely.

The locus of culture

In the previous chapter we examined culture in terms of three spheres: work, home, and public life. One of the most striking aspects of high technology is the possibility that these three spheres will be mixed as three trends merge: first, the trend towards renewed home work, both on the part of well-off managerial and professional workers, and on the part of often exploited manual and clerical workers in industries such as insurance and microelectronics; second, the trend towards portable work and communications facilities (portable personal computers, cellular radio-telephones, miniaturized stereo sets) and work in semipublic settings (the ubiquitous conference); third, the fuller invasion of a modified public space into home and office (increased political polls via telephone, telemarketing). In two different forms, homelessness may be a growing trend: one for the well-off, perpetually in motion; the other for the poor.

Identity, community and solidarity

In addition to the blurring of lines between different spheres of life, rapid transportation and telecommunications can be expected to continue to enlarge the share of the world's population included in the culture as a whole. Together, these two changes may pose sharp stresses in people's identities. Especially for those who are either economically or cognitively incapable of mastering the complex and overlapping structures of symbols required for full coping, life may be lived largely in a kind of anonymity in which machines will know one's name, but no one will be exactly sure of one's role in society. Those who are more successful at warding this off may do so by identifying with some small subset of humanity, connected through information technologies, who can form a kind of "microculture" with its own stability. Conversely, the people who succeed in maintaining a cultural identity in this way will also probably be economically successful; their work, while changing, will continue to involve imaginative manipulation of symbols. It is likely that the symbol-manipulating class as a whole will be able, through its microcultures, to maintain a complex social solidarity. The rest of society, including the poor and many manual workers, will find themselves increasingly deprived of the traditional means for social solidarity, or for organizing political opposition to an unfair status.

Because the microcultures will be partially impenetrable by outsiders, especially poorer outsiders, the chance for the sort of undifferentiated public space in which anyone may come to the attention of anyone else would be even smaller than now. Since it is in this space that people come to sympathetic or empathetic understandings of the lives of outwardly very different people, the chances for a humane and caring politics might diminish. On the other hand, such encounters sometimes serve to reinforce negative prejudices. Instead of prejudice, there may be something more like complete obliviousness.

The nature of reality and the issue of literacy

In a world where people may be expected to spend ever more of their time simply staring at a screen of some sort – a screen in which in many cases there will be an image the person can directly manipulate – the difference between image and reality may be less sharply apparent than it now seems to be, or than it was in the past.

Already, people confuse the media, such as television, with reality. Heavy television viewers, for example, see many "violent" acts every day. Most of them come to believe the world is a much more violent place than it "actually" is. Why is violence so common on television, and why do watchers come to believe it is real?

Our television system is set up on the basis of selling an audience to advertisers. To maintain the largest possible audience requires efficiently presenting images that people find it hard to turn away from. Violence involves rapid action and, since it promotes fear, is gripping; other kinds of rapid action, such as sports, work too, but once the convention for showing violence is well-developed script writers, directors, and actors find it hard to replace it by other sorts of equally gripping images. (If holding an audience were not a requisite, television might try to induce people to be excited about various ideas on which they then would want to reflect. But reflection means turning off the television, so it is to be avoided.) Therefore the easily comprehensible and well understood formulas for violence have a distinct advantage in holding attention. But this still does not answer the question as to why televised violence is taken to represent reality, rather than to be simply a highly conventionalized art form. Is television trusted because it involves such a life-like image? Perhaps, but another reason it is overly trusted may be somewhat different, having to do with literacy.

When a child learns to speak, he or she also discovers that it is possible to tell lies, to make up stories. Through that direct experience the child learns a certain degree of skepticism and mistrust that is an almost automatic part of gaining fluency in speaking the language. Likewise, fully to learn to read, one must also learn how to write, for then one can understand that a written or printed text carries no more authority or truth than spoken words. Only by trying it out does one come to realize clearly that writings can be mistaken or dishonest. (People who are uncomfortable with writing are more likely to believe that simply because "it is written," it must be the word of an infallible being.)

The essential problem with television may be that most of us do not know its syntax; we are partially "illiterate" in television because, despite years of exposure, we can only "read" and not "write"; our reading is complete since we understand hardly at all how mistakes, distortions, fictions, and lies may be introduced in images presented.

To put it another way, televised images seem as obdurate as reality because, like reality, we seem to be unable to change them purely at will. Command of the syntax of television that would come from ready access to production facilities at an age close to the time we begin to watch it would change that relationship. Similarly with the plethora of currently emerging information technologies: the culturally crucial question is whether we have an opportunity or capacity to participate. This is the reason that current movements to ensure "computer literacy" are important – not because we can all expect to get jobs as computer programmers.

If the coming of an information society is not to reduce human understanding and the cultural involvement of the vast majority of people, technologies to permit familiarity with the potentials of the media must be developed. In television, inexpensive cameras would have been of value thirty years ago. The centralized broadcasting model of the medium denied us those possibilities. The danger now is that only a few will gain the sort of access and familiarity all of us need. In addition, for full involvement, we will also need remedial methods in literacy, ways to translate the communications talents each of us possess into terms that will allow us to make sense of the proliferated variety of cultural signaling.

Lifecourse expectations

Information technologies now make it possible for corporations and other organizations to examine the profitability of their subunits, and to restructure these subunits in many ways, often without physically altering them. Subunits are likely to be opened, closed, bought, sold, or merged with great frequency. Permanence of employment in a stable occupation then becomes unlikely. Likewise, at least the more successful employees will be continually tempted to seek to further themselves by employing their talents elsewhere, perhaps in very different work. What people come to expect for the course of their lives will probably include rapid changes, moves to different locales, etc., creating enormous insecurity and instability for many. Even someone who may seem to be a highly successful person engaged in creative activities will have to worry about his or her capacity to continue to be creative to the same extent. Such developments will add to the fluidity of identity and sense of self, without necessarily offering a strong sense of community either. Openness to cultish religious groups, etc. may be a growing problem. The general question of what life is for and how people can maintain a sense of an ongoing rootedness in the world (and a stake in the collective future) may become sufficiently pressing as to pose dangers for the preservation of genuine democracy.

Flexible production, social stratification

For most of this century, the main highly mechanized form of production in the US was mass production – of very uniform goods – epitomized by ordinary electrical power. The reason for this was partially that the only high-productivity kinds of processes that were possible involved highly repetitive tasks. To do more would have required a capacity for changing the behavior of machines with great rapidity, according to need. That was just not possible. Perhaps it was not possible simply because the myth of social equality that had deep roots in this country made it seem desirable to industrialists that the same goods be produced for everyone, so that no effort was made at the time to develop alternative machinery.

In any case, mass-produced goods could find the largest possible market under exactly the assumption of basic equality. Especially in the post World War II period, that was the dominant myth in America. Every family would be part of the middle class, would own a house, drive a car, send their children to college, watch the three television networks, eat sliced white bread, etc. When the myth turned out to be only partially true – i.e., when the middle class discovered that blacks, Appalachian dwellers, etc. did not have equal shares – enough of a sense of scandal resulted to provoke many to strongly support reforms such as civil rights, food stamps, and the "War on Poverty".

But new technologies now permit efficient, flexible production. In many cases, there is no need to produce in advance of demand. Therefore, industrial growth no longer need be tied to enlarging the group of people who buy the same set of mass products. Instead, a smaller class can simply buy more and more of a highly varied set of goods, provided the income of this smaller class keeps rising. One possible basis for the income of this class to rise is that it itself produces not ordinary material goods but information, the value of which must be set more or less arbitrarily, and therefore can be continually increased relative to the output of more traditional workers. Signs that this is already happening may be found in the current distribution of wages in our society and the new importance attached to luxuries of all kinds. Real wages for many manual workers and for the poor are tending downwards, whereas, for a variety of college-educated workers, incomes are rising. The new luxuries appear to be associated with the lifestyles of the latter group.

If the trend continues, we could easily move towards a much more sharply divided and stratified society with two major classes: a class of highly educated creative workers – who will be making computer programs, designing clothes, directing television programs, producing the never-ending series of reports government and business seem to metabolize, etc.; and a class of poor, partially employed workers whose standard of living may begin to sink to the dangerously low level of the world's poor.

The alternative would be to deploy the flexibility that technology now permits for wider advantage. There is certainly no inherent barrier to doing so, but it would require modification of current trends in distributing goods and services, which would have to be based in turn on different standards of evaluating comparative prices for incommensurable goods, such as mass-produced objects versus information.

Work culture

Automation does not eliminate jobs altogether, because, at any given moment, certain activities are either not automatable or too expensive to automate. People have to do what machines cannot. Sometimes this involves fairly interesting work, but often it turns out to involve low-paying jobs where what the human provides to supplement the machine is very boring or very low status work. Even the interesting jobs are not skills in the traditional sense; many of them are also transitional because what cannot be automated now might be with advances in programming methods, computer modeling techniques, computer power in general, electronic sensing devices, and automation strategies.

This is not meant to imply that computers will necessarily ever be able to do everything that humans can. At present they are certainly extremely far from that; for instance, a computer coupled with television cameras is barely able to discern objects of standard, fairly simple shapes in a factory environment. Nonetheless, these limitations do not bear directly on the fraction of tasks that can be automated, because technology permits redefining tasks, for instance by redesigning the products computers might be involved in helping to produce.

The familiar example of automatic teller machines reveals something of what is involved. The machine certainly can't do everything a real teller can, but by standardizing some of the actions of the customers, the bank is able to use the machine to take over many of the functions of tellers. For instance, the customer now has to press a particular sequence of buttons to obtain balance information for a specific account. Before, there were many ways the customer might have asked the teller for this. The redesign of the process saved having to have a machine capable of understanding general spoken requests, a capability that does not now exist and is unlikely to be developed soon.

The limitations of automatic teller machines also leave some rather dull tasks to bank employees. One is examining the amounts on all checks deposited in the machine to verify the customers' deposit slips. Before, this was usually one of a series of varied tasks for tellers, but now it is more likely to be someone's full-time assignment.

The alternative possibility that new technologies leave open is to design the machinery, etc. specifically to create a set of interesting, fulfilling and open-ended jobs, jobs that involve social as well as technical aspects. A complementary way of dealing with automation is to alter our stance on the nature of work itself. We could automate those jobs that no one finds enjoyable. At that point the incentive to work would be in the character of the work itself – it would have to be interesting, worthwhile, and fun. Such work could also have its built-in rewards, in the sense of satisfaction in what has been accomplished and social gratitude for a job well done. Such a situation would allow the rather complete separation of jobs and income. With a society capable of supplying all the reasonable needs of all its members, working might not be a necessity for simple survival. These and other alternative possibilities discussed in this chapter will reappear in the form of more specific policy proposals in Chapter 8.

Gender and ethnicity

If new technologies make possible new divisions of social roles, then the advent of high technology could be an occasion to help end gender differences, as well as distinctions based upon ethnicity. However, that does not seem to be happening. From elementary schools to law offices, for example, computers are perceived to be essentially "male," even though in many cases the uses to which they are put are almost identical to the uses of primarily "female" tools such as word processors. Likewise, barriers of poverty and cultural difference have limited minority access to computers and related instruments. White males continue to write most of the computer programs, and as these programs become increasingly "user friendly," the users to which they are most friendly tend not surprisingly to have outlooks similar to those of the programmers. With enough conscious effort, overcoming these gender and ethnic biases would seem feasible. However, it is also important that gender divisions of society not be replaced by equally rigid class divisions. If professional women, for example, attain parity with professional men in the use of personal computers, it may still be that other women and men will end up permanently stuck with low status and low pay tasks such as home assembly ("stuffing") of electronic circuit boards or equally low-status childcare work.

Geographical relations and high technology

The effect of worldwide telecommunications on geographical relations was discussed in Chapter 3. What further changes seem likely?

As recently as 1970, an obvious answer would have been the incorporation of outer space into our ideas of place. But, if anything, the moon and the planets as real places seem to have left the popular imagination rather than entering it. For instance, ideas of permanent moon bases or colonies on Mars seem old hat, even though such things have never existed. Perhaps the basic reason is that places are of interest insofar as they are peopled, or at least could be peopled; if they have no recognizable culture, they should at least have a recognizable ecology.

Whether or not outer space has a future, current technological trends point to a very different kind of transformation of geography. Computers, television, and video games have begun to make a reality of what might be called conceptual space. A computer screen is quite regularly conceived of as a window opening onto a space, not all of which can be seen at once but around which the computer user can navigate by using a keyboard, "joystick," or a "mouse." Many of these spaces, at present, are visualized as two-dimensional – in essence, large, imaginary pieces of paper. But, in principle, three or even more dimensions may be included, and these may come to represent an increasing portion of computer users' sense of the world. Since different users will be able to move in the "same" spaces, and there communicate with each other, as is possible in data banks, electronic "bulletin boards," and other computer networks, these are not so much "inner" spaces as alternative spaces of human relationships – a real addition to geography.

For everyone who has access to this conceptual geography and for whom moving around or building structures in it can be meaningful, the possibility of plenitude as opposed to scarcity as the fundamental reality of human existence would at least begin to make sense. Increasingly complex and "real" human experiences would be less and less subject to resource limitations; within this realm, at least, there could certainly be enough. Even as metaphor, the possibilities of plenitude within the conceptual geography mean that, within some very broad limits, it is possible to envision a society in which a wide variety of satisfactions would be available to everyone without destruction of the natural environment.

However, current trends do not suggest that our society is heading in those directions, just as the conceptual geography is not being made equally accessible to all, but rather only to a portion of the upper middle class. One of the goals of a sound technology policy would be to enlarge such access and to specify more fully, and less metaphorically, just what the possibilities and limits are.

Conclusion: uncertainties and hopes

The preceding chapter ended with a query: can high technology fulfill the hopes vested in it as a means to reestablish an economic and social stability? In other words, is there a viable social structure that can be associated with the set of new technologies, and which will leave intact the basic institutions of mid-twentieth-century America?

A viable social structure should allow almost everyone to have a social role and to perceive that role as worthwhile. With that stipulation, a tentative answer garnered from glimpsing the various facets of life just presented is "No." On the other hand, the same glimpses suggest that, with some institutional changes, which would also change the direction of technology, a better and more viable way of life would indeed be possible.

Part Two of this book can be read as an argument for the latter claim, including in some detail just what the institutional changes might amount to. A moderately convincing argument for the former point would probably require an equally lengthy exposition. That would be out of place in this work, and instead, what is offered below is intended as a brief and impressionistic substitute.

The combination of social forms, purposes, and daily patterns that were "the American way of life" cannot be reassembled with a high tech gloss. That impossibility is summed up in the recent history of the standard-bearer of the old life – the General Motors Corporation. In order to remain in tune with the times, GM has recently purchased two smaller giants, Hughes Aircraft, a major producer of satellites and high-technology weapons, and EDS, an electronic data processing corporation; in addition, GM has invested heavily in a number of small robotics firms.

GM's ostensible reason for all this is to be able to continue to be preeminent in the production of automobiles – the new, electronics-laden cars to be built in the most modern of plants. At the same time, hedging its bets, GM is also to be "the automation company," providing the means for automation to many other industries.

If the suggestion in Chapter 3 is correct, General Motors is caught in a very real dilemma: the new technologies cannot simply be new ways of doing more or less the same old things, because then there will not be enough of a change in the overall industrial and economic structure to overcome the difficulties that make transition to a new set of technologies desirable in the first place.

The automobile is an old technology; GM is proposing to update it by building in all sorts of electronics. This may have the effect of altering production patterns, but for the most part, it will not change the uses of the auto or the way of life that goes with having one. (Contrast this to the changes that went with the transition from rail to automobile.)

Adding electronics to the car is, in effect, a kind of window-dressing, not a contribution to the transformation to a new stability that GM obviously wants to be part of. GM's hedging of its bets is a way of recognizing the limited future of the old mass-consumer culture. While it could have invested its resources in any other industry at all, its choice of a new direction is a striking confirmation of the difficulties it and other corporations face.

One of GM's chosen alternatives, automation, is not to be sold to private individuals, but to other large manufacturers. What will these other manufacturers sell? Like General Motors, other major manufacturers in all advanced countries are eyeing the defense and capital goods markets, if they are not heavily involved in them already.

Giant corporations, in general, are also rushing to increase their ownership of information, especially in the form of research capacities and results. It seems to be the only chance they have of maintaining supremacy. But these forms of information can be property precisely because they act as templates – for copies of the information, for new computer-run machinery, for new production processes. They all have the character of being capital goods for making capital goods. In other words, the new corporate holdings constitute an exponential increase in productive capacities.

In the past, the imperative of corporate growth always meant that production capacities had to be increased until they were more than the market could bear. At that point, maintaining a sufficient profit margin required cutting back production, leaving excess capacity – just as OPEC's survival hinges on cutting back oil supplies. Of course, that cutback did not mean that all needs had been satisfied, but rather that everyone was served who could pay the minimum price the corporations were willing to charge. Had all the excess capacity been utilized, prices would have fallen, and more people's needs could have been satisfied.

The new forms of capital increase the tension between the imperatives of corporate growth and filling needs. The information that is the capital could be of use to society without the corporation's having to do more than let it be used; to the extent it could fill needs at all, there is no limit to how many people that same information could help. But corporate survival requires that this not be permitted; there

would have been no point in assembling the information –
whatever it may be – in the first place, were it just to be
released freely. Think of this in industrial capacity terms. If,
before, industry was operating at three-quarters of capacity,
maximum production would have provided one-third more
products. But now, each company restricts to itself great
volumes of information that could be utilized everywhere
on the planet for producing goods or services of some kind.
That a lot of that information may not have very good uses
is beside the point. If it has any uses, its restriction amounts
to having industry work at a tiny fraction of capacity. (Were
research and development focused exclusively on worthwhile ends, the benefits that would potentially be available
would be all that much greater.)

Beyond the repressive practices each company must instigate to preserve its intellectual property, the unused industrial capacity is in itself a growing absurdity. The industrial
system has become a system that can thrive only by
increasing the gap between promise and performance. To
proceed along these lines, companies have to view each
other as virtually the only possible market – so that in the
future they will be able to produce even more to sell to each
other.

An economy dominated by capital goods can hardly function in the way the American economy did in the past, for
then such goods were bought largely with the – understandable – intention of producing consumer goods with them.
But, within the present system, a sufficient increase in
consumption capacities appears hardly possible, so there is
little point in producing the consumer goods the new capital
goods would make possible. The question remains, what
are these new production capacities to be used for?

If the epitome of the old way of life was the private
automobile, the best candidate for epitomizing the new is
apparently the personal computer. The difference is that,
whereas the automobile was a consumption item that facilitated both other forms of consumption and the necessary
travel between home and work, the personal computer, as

it has been developed, turns out be much more a tool for production – a management tool, a means for investment, or a means to produce texts for subsequent publication – among other productive uses. If the problem underlying the hope for a technological transition is surplus productive capacity, yet again part of the solution – a new class of consumer goods – turns out to be still further surplus capacity.

Apart from each other, the main available new market for GM and its compères is the sphere of national security, which matches all corporations as an instigator of new technology, and more than matches them in terms of perversity. The "Star Wars" effort represents a trend comparable to industrial trends. It formalizes the arms race as a race of research towards ever-new classes of weapons, so that what had previously amounted to expanding capacity for destruction now becomes expanding capacity to expand destructive capacity – obviously, without a parallel increase in things that can be destroyed with any equanimity.

Not only does the GM example serve as a symbol of a very odd sort of change, but the oddness suggests that this giant firm, capable of affording the best in strategic investment advice, may nonetheless be mystified about the future it is so proudly and urgently attempting to shape. Can it be that the most powerful forces of the past – and indeed of the present – are flailing out desperately to try to maintain control without being at all sure how? That while firmly embracing the model of technological transition with one arm, they are trying to ward off its consequences with the other?

A desperate attempt of an aging giant to hold on to power does not necessarily presage a better future. Without a clear idea of who might gain that power instead, we have no reason to be optimistic. And, in any case, in the giant's flailing, great damage might be done.

But, if even GM is uncertain, it is just possible that the nature of the transition is still not fixed. Part Two will attempt to spell out that an information-centered society need not be dominated by only a few, need not be power-mad and war-crazed, and can permit a new openness, a new abundance, a new democracy. That will require new directions for technology, which, in turn, will necessitate political change, as the next two chapters will clarify.

5 The role of government technology policy

If technology's impacts are vitally important, and if those impacts depend on the values and interests that inform the development of technology, then how are those values and interests to be changed? If the growth of science and technology demands a restructuring of relationships among the institutions of society, then how can that restructuring be accomplished? The most obvious answer to questions like these is that such changes must occur through the political process, and that the mediating influence of government must be brought to bear in new ways. That answer leads to a new question: what are the ways in which government has influenced technology up to now, and what are the interests and values represented in that influence? This chapter provides a brief answer with respect to our federal government – certainly the most powerful actor in the field of technology.

The federal government is not at all of a piece; it consists of many different institutions, ranging from the legal system to highly specialized, goal-directed agencies. These different institutions can work very differently and, at any one time, proceed with different ends according to sharply different motives. Furthermore, throughout US history, the range of institutions and the central values of government have altered repeatedly, and can do so again.

Some of the factors that influence a particular agency are its explicit purpose, its special constituencies, including its own staff and its contractors, ideas in the air at a given time, and the ideas of the political party in power. All of these matter; even though a particular administration may try to redirect government, an agency set up with a definite function is likely to remain at least partly attuned to that function and the values it implies. The agency almost certainly will be closer to those values than an agency whose function is stated in different terms. Even under the Reagan administration, the Equal Employment Opportunity Commission is more likely to be attuned to the issue of minority employment than is the Pentagon. One way, then, to gauge the values that shape technology policy is simply to examine the designated purposes and the constituencies of the primary agencies having some form of authority in this area.

An agency's constituencies include its own staff and all those groups who must develop special relations with it, as well as the constituencies it supposedly serves by dint of its stated function. Thus, the Pentagon's constituencies include not only the military, but also defense industries and the general scientific and technological community, since the Defense Department funds so much of their work. To a lesser extent, through its wider economic impact, as well as its supposed defense function, the Pentagon could be said to have at least the prosperous, throughout the country, as its constituency. Other agencies, such as the Department of Labor, have special constituencies, such as labor unions, to a degree that other agencies do not.

Agencies are also influenced by ideas with current support. During the 1970s, when concern for the environment received government backing, the Air Force proposed a solar power project as part of the "race track" plan for siting MX missiles in the Utah desert. Currently, industrial competitiveness with other countries is a major concern; so interest in increasing productivity – especially through automation – is widespread among government agencies.

Research, development and procurement policies

The federal government influences and reacts to technology in three broad ways:

1 It directly influences the course of science and technology by sponsoring research and development, and almost as directly through its procurement policies. With the possible – although unlikely – exception of the Soviet government, the US government is the largest single direct sponsor of research and development, and thus exerts a substantial direct influence on technology in the world as a whole.

2 It strongly affects the overall climate for technological development in the private and nonfederal sectors, via patent and trade policies, tax policies, regulation, court actions, and by other means. Because many of these policies are implicitly international in scope, the effect is again very wide.

3 It reacts to the impact of technology through a number of routes, which then result in laws, regulations, and sometimes even reorganization at the agency, department, or – occasionally – government-wide level, as well as through changing international regulations. The government itself becomes more technologized, both in its mode of doing business and in its outlook. Decisions with genuine political content can be disguised, whether by intention or not, as either technological choices or as the unavoidable outcomes of purely technical considerations.

The most striking aspect of these activities is how unequally they are spread through government agencies. Agencies concerned with the interests of workers as such, of consumers, of the poor, or with community life, equal rights, or democratic involvement have almost no funds for research and development, outside the area of health and safety. Even there, development funds are quite limited.

Few will be surprised that by far the greatest share of research, development, and procurement capacities is under the control of the Pentagon. In fiscal 1984 the estimated Pentagon R&D budget amounted to 66 per cent of the total federal R&D budget, or some $27 billion. Of this sum, $23.5 billion were devoted to development – some 87 per cent of all the federal government spends towards actual incubation of new products and processes. Substantial additional defense-related development funds are to be found in the Department of Energy nuclear weapons budget and in the NASA budget. The small remaining fraction of development funds is spent largely for air traffic control, increasing productivity (by both the Commerce Department and the Agriculture Department) and to a very limited extent for health care.

By contrast, departments and agencies more focussed on human needs or the needs of the less well-off have hardly any development funds at their disposal – especially outside the health care field. For example, the Department of Labor, the Civil Rights Commission, the Department of Housing and Urban Development, the Human Services part of the Health and Human Services Department all have essentially no money for technological innovation. Barred from direct influence, these agencies have tended to view technology as a given (an attitude the Pentagon would never take). They have therefore remained unable to shape technology according to the values and interests they are supposed to be serving.

Basic and applied research budgets are not quite so lopsided. Still, such budgets for the human-centered agencies mentioned above are minuscule. Since the federal R&D establishment has considerable influence in determining the areas of pure and applied research that agencies such as the National Science Foundation support, the priorities of development funding weigh heavily in these allocations as well. There is nothing conspiratorial in this; it is simply that knowing what they are trying to develop, Pentagon technologists have a clear idea of what applied research they need, and the applied researchers, in turn, know what basic research they would like to see done. Thus, even for "pure" research, the Pentagon's goals are of far more significance than any domestic social program. Since existing basic and applied scientific knowledge in turn sets limits on what technology can be developed, the wrong values become ever more deeply embedded in what seems to be possible.

Because the US defense R&D budget is such a large proportion of the total of all world R&D, it influences the direction of technology in many areas. Also, the large latitude the Pentagon has in deciding what to fund has permitted it to influence technology even in areas only remotely connected with weapons. One example is the development of "CAD/CAM" (computer-aided design and manufacturing) technology for automating production in the metal working and related industries (including automobile, aviation, heavy machinery, and household appliances) – all with little public debate or awareness. Workers, who have most at stake in such research, have even less chance to influence or even to know about it than they would if the government were to set up an agency explicitly to develop and spread automation.

To see the values behind its technology policies, we have to look at the purposes of the military. These are, of course, complex. There is reason to believe that the vast expansion in peacetime military expenditure that occurred at the end of World War II was connected in large measure with preventing not another war but another Depression. Keynesian economic principles led to viewing Pentagon staffing and procurement policies as a vehicle for economic growth.

It makes a certain kind of sense for defense expenditures to be the primary means of increasing demand in order to keep the capitalistic economy functioning. Simply put, the military does not duplicate any private sector functions. It follows that all of military spending may be viewed as pure gravy, in no way competing for markets with any profitmaking enterprise in any sector. Other prospective ways to enlarge the economy that have this feature are hard to find, and when they exist, may seem harder to justify than the military. For example, a guaranteed income would put additional purchasing power in the hands of the poor and unemployed, thereby increasing the overall level of economic demand; however, not only would this diminish the necessity of working for an adequate living, thereby possibly undermining the work ethic and endangering the ability of employers to find willing workers, but it would compete with the private pension and financial sectors even more than current Social Security does. Because there is no private sector activity claiming to defend against foreign threats, military spending is perfect for the Keynesian role. In this role, it functions as pure waste of funds.

Secondly, of course, the military has the function of maintaining or enlarging American power in the world. The power to destroy, which is the major power the military has to offer, is not terribly efficacious in controlling the modern world. For instance, America's most significant challenger is probably Japan and not the Soviet Union, but the American military already can be said to occupy Japan, without much effect on Japanese policies. While military action could lead to the economic and social destruction of Japan, it is not clear how, even if we wanted to do so, we could use military force to get a strong Japanese economy to open itself more to American goods. Military force simply turns out to be too blunt an instrument for the real concerns of the modern world. Nonetheless, it still provides the illusion of power, and the more helpless and hapless the US feels itself to be in other respects, the more tempted it is to try to further augment military force.

(Perhaps it is worth recalling here that if the US, like other national governments, does find itself increasingly, as Richard Nixon feared, "a pitiful, helpless giant," then the impacts of technologies intended to have the exact opposite effect are one reason. The power of technology, as argued in Chapter 2, is universalizable, and therefore attempts to use it competitively must ultimately backfire – in a world where no one nation is a majority.)

The third function of the Pentagon is to serve its own constituencies, and in doing so to maintain the values of the military. These values include:

- an underlying assumption of international conflict, conquest and control;
- faith in the power of destructive force;
- constant readiness and vigilance – which requires an infallible communications and intelligence gathering system, classification of secrets, mistrust of civilians, etc.;
- maintaining a hierarchical chain of command – including observing the social separateness of an officer corps;
- opposition to insubordination;
- avoiding battle casualties, when possible, while inflicting them on enemies (this value clearly leads to interests in remote control, automation of weapons, etc.);
- arms races as themselves a means for obtaining superiority.

Technology in support of these values is often arcane in relation to the needs of a world at peace. The various military forces are interested in such things as communications at extremely low frequencies as a way of maintaining contact with submarines during a nuclear war; they need planes and missiles undetectable by radar (stealth technology, involving materials, aircraft shapes, and even paints of little likely civilian use.) Neutron bombs, high speed projectiles, "daisy cutter fragmentation" bombs, helicopter gunships, intercontinental missile guidance systems accurate to within a few hundred meters, along with most other items on their incredible shopping list, are all technologies of military significance only.

But there are many other instances in which Pentagon-sponsored research does have likely civilian applications. Indeed, this fact is often offered to justify military spending – as if it were impossible to do research directly for civilian ends, and as if military sponsorship did not lead to distortions and did not hide from the public eye just what directions of technology are funded. We have already seen in the case of CAD/CAM that there are serious problems in allowing military dominance of a technology with vast possible civilian uses. The field of "artificial intelligence" is another striking example. This is a broad area of research

– now almost entirely Pentagon-funded – aimed at enabling computers to perceive or reason as much as possible in the manner that humans do. Its potential applications for the military range from so-called "smart" weapons, to taking control of a nuclear war, to eavesdropping on telephone calls for spying and counterspying. The field also involves a deep and intense debate as to how much computers may be made to think and perceive as do humans, and the Air Force has gone so far as to sponsor a Heideggerian philosopher's attempts to show that they cannot. Whatever the outcome, this is clearly a debate of vast human significance, but even more important is another question that simply does not get raised in any serious way under Pentagon purview. And that is: Should computers be made as much like humans as possible? Do we want a world of machine spies, of machines capable of replacing humans at many jobs when world-wide unemployment is already so high, or even of machines similar enough to humans to be capable, say, of suffering? It is arguable that there is nothing to worry about as far as that last question is concerned, but the others are clearly germane at whatever level of artificial "intelligence" has already been achieved. As long as the Pentagon can hand out money in virtual secrecy, while other agencies cannot decide research directions, we are leaving this important public debate to the generals and their civilian allies. There are numerous other areas, from new clothing to flight control to emergency medicine, where military values determine civilian technologies without adequate debate.

To sum up, the technologies of interest to the Pentagon can be classified in two groups, depending on whether or not they are likely to have nonmilitary applications. Both groups are problematic. When there are nonmilitary applications, the influence of military values on development will have shaped the basic technologies in ways that are unlikely to accord with democratic values. On the other hand, when technologies are solely of value to the military, the engineers and scientists involved in their development have a special interest in perpetuating and accelerating the arms race.

While it is difficult to see how such a race can really augment American power, it is obvious that the continuous proliferation of new weapons systems of all sorts complicates peaceful relations and comes to dominate foreign policy, promoting tensions that can lead to war. It is also clear that the greater complexity involved with an ever expanding collection of military technologies increases the variety of mistakes, accidents, miscalculations, or sabotage both destructive in themselves and potentially capable of precipitating full-scale war.

The giant scale of the Pentagon procurement budget, and its enormous influence both within and outside the government, extends a penumbra over technology even beyond its immediate reach. Numerous companies without their own defense contracts align their innovative efforts with one eye on military requirements. Other government agencies, such as the Energy Department or NASA, are in part servants of the Defense Department. DOE, the current reincarnation of the Atomic Energy Commission, originally was placed in charge of nuclear development in order to prevent direct military control. Instead of decreasing military power, the result is that nuclear weapons are designed to military specifications in DOE labs. The nonmilitary projects of these labs, then, often turn out to have military overtones as well. For NASA, giant projects such as the space shuttle are intended to serve military as well as civilian purposes and are designed with that in mind. One of NASA's "civilian purposes," as evidenced so clearly in the race to the moon, anyway seems to be the quasi-military goal of demonstrating American might.

Before leaving the field of military research and development, it is worth dealing with the commonly held view that military research is not in itself dangerous unless it is coupled with full development and deployment of weapons. In the existing values framework, weapons-related research is the first step down a slippery slope to deploying a weapon. Once the research is done, if the weapon seems feasible, it can then be argued that if we don't develop the weapon the other side will. People who have already devoted a substantial portion of their careers to the research effort see their own vindication and further advancement as likely to arise from development and deployment. Thermonuclear weapons were promulgated with special zeal from the late forties on by the person who had been assigned to work on the possibility as a sidelight of the Manhattan Project of the early forties. (The Manhattan Project developed ordinary nuclear weapons – the atomic bomb – and there is evidence that a number of leaders of the project were quite eager to have the handiwork demonstrated to the world by being dropped on Japanese cities.) The same story has been repeated with neutron bombs, a variety of missiles, laser weapons, and many others.

There is a version of the argument that weapons research itself is safe: such research is necessary to prevent "techno-logical surprise." In practical terms, "preventing techno-logical surprise" means making sure the Russians cannot threaten us with a weapon for which we have no counter. But how can such research prevent us from being surprised? Research depends on inventiveness, creativity, and hundreds of choices as to what direction to pursue. There-fore research conducted in secret could still lead to surprises unless we pursue an infinite number of different lines of research. The actual effect of the notion of preventing tech-nological surprise is therefore greatly to increase the amount of weapons research that can be justified, and so increasing the possibility that we ourselves will develop a weapon that will later be copied by the Russians and end up being dangerous to us.

When we recognize that research is never conducted in complete secrecy, the situation does not improve. If we are doing research to help us follow what they might be doing, we are simultaneously giving them numerous ideas they might otherwise not have had. At the same time, neither side is actually able to assure itself that the other side isn't making advances it hasn't thought of – and won't think of. There is no technological way to prevent that; only a peaceful world can do so. An effort to develop technologies that could really help promote peace, by equalizing living conditions, aiding understanding and contacts between nations, reducing secrecy, and lessening the opportunity for one country to gain by having power over another – these efforts, difficult as they might be, would be far more useful than continuing research into new weapons to prevent surprises.

Impact of government funding of basic research

Government development efforts are concentrated in the military sector substantially for the same reason that military spending as a whole is the uncontested area for Keynesian waste: avoiding competition with private sector initiatives. These amount to the initiatives of the large corporations and a very few others.

Scientific research – especially basic research – is different. It is usually only very remotely connected with any particular profitable applications; therefore no particular private corporation can be expected to fund it. Since it supposedly helps industry in general, such research is understood to be a legitimate activity of the federal government. Yet the more basic the science, the more it needs interpretation to be applicable for any purpose. In effect, that interpretation is what is called development. To undertake it requires the capability of understanding the scientific literature, the ability to reproduce experimental set-ups in development laboratories, and the capability of translating such work into specific processes or products. With the complexity of current technology, these requirements imply large staffs, well-equipped, expensive laboratories, and substantial facilities to test and build models. In short, most development is costly. If government doesn't supply the funds, only well-heeled corporations can. Consequently, the basic research government does fund is of far greater help to large corporations, and to a few, middle-size high technology firms, than it is to smaller businesses, communities, other institutions, or workers. The policy of concentrating on funding military development sharply increases the power of large corporations as well.

Peculiarly, the waste and folly associated with government sponsorship of weapons technology has been used as an argument to keep government out of funding other kinds of development. Government is inherently so wasteful, the argument runs, that products it develops could never be sold. Furthermore, government decisions on what to develop are assumed to be inimical to freedom of choice. If private companies offer new products on the market, then consumer response can show them in what directions further development should go. According to this argument, the market acts as the most effective means of popular control of development, and it is one that the government can only damage by interference.

There is some truth, in certain domains, to the notion that the market permits choices among different technologies, but there are also numerous flaws in this position. As we have just seen, development involves interpretation of discoveries sponsored by government, and this interpretation can usually only be carried out by companies of substantial size. Because they are effectively the clientele for federal research programs, they directly and indirectly influence the areas to be researched. For instance, the Agriculture Department's model of what is worth pursuing accords with that of the giant agribusiness corporations, as well as with that of the farm machinery, fertilizer, pesticide, animal breeding, and hybrid seed industries. Likewise, the National Institutes of Health are oriented to a concept of health care centering on hospital care and prescription drugs. In both cases, the large nongovernmental institutions – mostly corporations – that influence the government also influence the market through advertising and through the government as well. Advice to farmers offered by the Agriculture Department is tied to the concept of high input agriculture. Medical programs such as Medicare are equally tied to cure, rather than prevention, as the basic model of health care.

The very absence of government support for development obviously aligns the market in certain directions; freedom of choice is undercut and not enhanced by the hands-off policy supposedly in effect.

Even if there were some way for the government to be neutral with regard to the direction of technology, the market would be an inadequate means to assure that many important needs were addressed. First, there are those groups who have little direct power to influence what is produced because they do not control purchases. The poor, simply because they are poor, are in this category. Workers and their organizations have very limited legal rights to influence purchasing decisions relating to workplace technologies; therefore, ordinary workers also cannot affect developments of great importance to them. A second weakness of the market results from the fact that purchases tend to be individual decisions; the needs that exist only on the level of the community as a whole tend to be undervalued. These include environmental needs, such as quiet, that can only in part be met through regulation. There are also needs for social cohesiveness, for example, that regulation alone can do very little to assure.

A third category for which the market fails is that of special requirements of people with average incomes. Technologies of production influence the range of variation of goods offered on the market. Often this range is inadequate for many people, who nonetheless constitute only a minority on the market. For instance, most shoes are no longer available for people with very narrow or wide feet. This is because shoe manufacturers find it more profitable to produce many styles than many sizes. Since, for each style, additional sizes require additional lasts, more complex record keeping, larger storage space for inventories, etc., this choice is directly related to some rather simple technological choices. As almost everyone is in several minorities of this type, everyone suffers from this effective limitation of the market.

Finally, there are very many potentially valuable goods and services that would be expensive to develop but that most potential users have probably never thought of. Goods of this type sometimes are developed, as an act of faith on the part of venture capitalists or other entrepreneurs, but which ones are developed remains arbitrary, with the only overall influence being the values venture capitalists have in common. Without some public, dialogical process, many feasible and much needed developments will be permanently blocked.

Presiding over private sector innovation

If the government leaves most innovation to the private sector, it nonetheless influences the pattern of that innovation by a whole range of services, regulations, and laws. Arguably the most extensive such influence is the regulation of who can make what, through the intellectual property laws: patent, copyright, and trade secrecy. Other important activities include establishing and maintaining standards of measurement and compatibility for many industries; regulating on the basis of factors such as safety, health, and environmental protection, and extending property rights (e.g., assigning radio bandwidths to particular broadcasters). Only very rarely do such actions explicitly promote democratic values.

Patent and copyright laws were originally permitted by the Constitution for the social benefit that presumably would result: "The Congress shall have power . . . [t]o promote the progress of science and useful arts, by securing for limited times to authors and inventors the exclusive rights to their respective writings and discoveries," states Article I, Section 8. Despite this background, ownership of patents and copyrights is widely understood today to be purely a right – one accorded not so much to inventors and authors as to their employers and publishers. Hardly anyone appears to be much concerned with whether these rights really are either necessary or sufficient as incentives to innovation, or whether the innovations related to them are valuable to society as a whole. What makes this state of affairs particularly remarkable is that, in a period when "deregulation" is high on the political agenda, in effect one of the largest – if not the largest – and most arbitrary systems of regulating private behavior remains unchallenged.

As an incentive to innovate, the intellectual property laws leave much to be desired:

1 As they currently stand, the patent laws favor the wealthy – usually wealthy corporations. This is because exactly what a patent claim covers may be murky, so that there is usually latitude for legal interpretation; the side with more money can wage the legal battle to the point of exhausting the other side. In addition, a patent is useless without the capacity to exploit it. Only inventions that will attract development and marketing funds are worth patenting at all. What is encouraged is not invention per se, but a class of inventions of interest to corporations. Corporations may vary considerably in their interests and business approaches. Still, large corporations are not the same as people; their interests are not the same either. The patent system helps bias inventions away from the interests people do not share with corporations.

2 Likewise, there are many conceivable inventions that are not patentable, perhaps because they are too general in concept. Existing inventions obviously too broad to patent include the basic ideas of the personal computer, abstract art, free verse, soil conservation, chocolate chip cookies, assembly lines, inflight movies, synthetic fibers, etc. If the patent system is to act as an incentive, the originators of such ideas should certainly be rewarded in some way. To have them benefit by owning total rights to these broad ideas hardly seems desirable.

Meanwhile, the patent system may well be focussing technological effort on a narrow, and not necessarily socially beneficial class of inventions. For example, once a successful invention reaches the market, the patent system encourages would-be competitors to come up with technological alternatives that perform the same function while circumventing the patent. In many cases, this leads not only to a nonfunctional duplication of effort, but to a useless complicating of life for buyers of differing versions of what is essentially the same product. Repairs, replacements, or shared use are all complicated by needless diversity of parts and methods of use.

3 The difficulty just discussed is one way in which the patent system leads to higher prices. A majority of commerce now involves patented items, so that direct price competition between identical goods is mostly a myth. Since patents allow a monopoly, whenever the invention is especially useful and not easily duplicable, the patent system automatically helps promote an unequal society. Innovators are at times able to benefit out of all proportion to their effort, while lower income groups have still less chance than they would otherwise to afford goods that might be especially important to them.

An even more glaring inequity the system encourages is patenting to block competition, thus assuring further monopoly. Most very large corporations indulge in this practice to some extent. A firm patents what may be a considerable innovation, with no intention of producing it, simply in order not to have to risk that a competitor will develop it and thereby endanger the first firm's command of an existing market.

A similar inequity is that a few large corporations can agree to exchange among themselves the licenses to a wide pool of the patents they hold while excluding outsiders. In this way they can make an entire technology more or less their exclusive property.

4 Current law forbids patents on what is common knowledge; if an inventor has already published so much as the broad outlines of the invention, a patent can be denied. This enforces secrecy in fields of research that may result in

patents. The recent surge of interest in "genetic engineering" as a growth industry has led to a significant reduction in the free flow of information in the underlying science of molecular biology. This trend not only inhibits potentially worthwhile research, but lowers the possibility of informed public debate. It means that large corporations which can individually afford to support substantial research communities in the field will gain increasing power to determine the direction and scope of technology and even of the underlying science. Licensing or trade secret pools, as already mentioned, can further strengthen a few firms' domination of the innovative process.

5 In addition to the constitutionally mandated "intellectual property" registration system, there has always been a second system – trade secrecy. Trade secrecy may be traced back through English common law to extremely ancient practices by guilds and other groups of maintaining their "arts" in secrecy as long as possible – even, at least in principle, forever. Originally, trade secrets were maintained through solemn oaths, with the punishment for breaking these oaths presumably left to the groups who extracted them in the first place. Today, however, trade secrets are enforced through the court system, often by means of contract law. Technical personnel typically sign a trade secrecy agreement upon being hired by a corporation. Even if they subsequently change employer, they are generally bound not to reveal technical secrets learned in their work, including those they personally originated. In practice, court enforcement of these contracts makes it a form of prior restraint, heavily eroding freedom of expression, and denying to the public an understanding of what is possible and of technologies that already exist. The system helps to further narrow the possibility of public control over the direction of technology.

Tax policy

The climate for innovation and for the adoption of technologies is affected by tax policy. There are two main ways
in which Congress has sought to influence technological
decisions in recent years. The first is the research tax credit
for up to 25 per cent of research expenditures; the second
is depreciation schedules that allow new machinery and
other equipment costs to be written off in as little as three
years. As in the case of patents, it is difficult to tell whether
these policies directly affect corporate decisions, but
indirectly they do put additional financial leverage in the
hands of those managements oriented towards innovation,
or at least towards buying new equipment and supporting
research. The tax law, again like the patent law, affects the
direction of technology by omission rather than by
mandating any specific benefits. That is to say, the only
values obviously supported by these laws are profit and
newness, not any other social benefit. Without any requirement, or even any inducement, to move changes in socially
beneficial directions, social harms – such as the automation
of desirable jobs – are inevitable.

A host of similar governmental activities support values
such as competitiveness and profitability, or even encourage
a commodified approach to areas that do not need to be
commodified. The National Technical Information Service
provides information obtained at taxpayer expense for sale,
thereby limiting access; access is further limited in practice
by the high degree of expertise required to make use of the
information in the form in which it is supplied.

Regulating for equality

There do remain a few regulations that support equality of
access to technology on a limited basis. This is particularly
notable in the case of communications – both telephone and
broadcast. Minimal local telephone service at low cost is

currently under threat, but still survives. Television broadcasters still are required to meet some standards of service to the communities they broadcast to, and there is still a fairness doctrine insuring certain contested views equal access to programming, as opposed to commercial time. The Rural Electrification Administration, and the Rural Telephone Bank, along with the Tennessee Valley Authority, still exist in part to assure a form of technological development for the rural poor.

Policy making and oversight

There are a surprising number of entities that appear to be charged with overseeing technology and examining its effects. They include: the National Academy of Sciences, founded to advise the government during the Civil War; the President's Science and Technology Advisor; the National Science Board; the Patent Office's Office of Technology Assessment and Forecast; interagency coordinating committees; the Congressional Office of Technology Assessment; and the House Committee on Science and Technology. The very multiplicity of these groups helps all of them avoid the difficult and important questions, while maintaining the impression that all bases have been covered. Virtually all seem to accept without criticism the model of growth and international competition that already dominate policies. Many of the advisory panels are limited to offering "purely" technological or scientific advice, hiding the values with which such advice is necessarily tinged. In many cases, the advisors have direct and long-standing connections to the government programs or corporate practices they supposedly are assessing.

Probably the best such agency is the (Congressional) Office of Technology Assessment; but even its studies suffer from being far too short range, from taking for granted, for instance, that the current relationship between corporations and workers is the correct one, and that values which underlie programs such as the military's are not to be questioned, even when they impinge on many important social concerns.

Even more than most government activities, technological projects are future-oriented. Most programs exist to underwrite or promote innovation. Yet attempts to assess the future consequences of technology are rare. One might suppose that avoiding such assessments is only prudent. Since the innovations haven't yet occurred, we don't know what they will really amount to. We also don't know how they will interact with other future developments, technological or not. In short, there are enough uncertainties that accurately predicting the future is not merely difficult but impossible. However, were this to be accepted as the reason not to make assessments about the future, it would have to be followed consistently: no long-range future planning would make sense. That certainly is not how the government operates.

The Pentagon is already seriously concerned with aircraft for the year 2000. The controlled nuclear fusion program is unlikely to yield any public benefits until 2020, at the very earliest – assuming it will ever be beneficial. At that point the program will be close to 70 years old. If actual government research and development programs are intended to shape reality several decades hence, it is clearly perverse that projections of likely social consequences are made, with any seriousness, only for a much shorter range. The obvious result is that desirable social goals don't enter technological plans in time to have much effect.

The opposite mistake is also common: technology is ignored in social policy. At the start of the Reagan administration, there arose a substantial national debate on the future solvency of the Social Security system. The issue was not the next few years, but periods as remote as 2030. Projections to that date, roughly the retirement year of today's youngest workers, were used as arguments about present payment policies. Yet one factor that would surely make a difference just as great as levels of monetary income to future retirees is how future technologies could affect standards of living.

While the future is unknowable in detail, it is possible to consider how current trends might proceed, where some of these trends will clash, and what sort of choices might make some difference. For technology policy to be sensible, such activities, though difficult, are essential. Of necessity, these activities would require considerable effort; the full-time work of a few hundred to a few thousand people might be necessary. That scale of effort simply does not happen, either in this country or elsewhere, whether in universities, corporations, unions, foundations, or the government. None of the existing government oversight agencies has the resources and independence to undertake or even to sponsor such an effort.

Technology as politics

Politicians and bureaucrats often are eager to technicalize a decision so that political passions can be sidestepped. Clearly, democracy is also sidestepped in the process. There are also occasions when technical professionals believe that, by couching a political argument in technical terms, they have an entrée, which would otherwise not exist, into the debate on an issue they care about. Debates about weapons systems often take the form, not of whether the weapon is justifiable in terms of the proper and legitimate position of the US in the world, but upon whether that weapon will "work." The debate is further complicated because it involves two different meanings of "work:" for something to work in a technical sense is not the same as working in a social sense. Contraceptives may prevent conception very well under controlled conditions, but the existence of contraceptives has not eliminated unwanted pregnancies. J. Robert Oppenheimer, the director of the atomic bomb effort, later opposed developing the hydrogen bomb on essentially moral grounds. Who was in a better position than he to have reflected on the moral consequences of developing weapons? Yet he couched his opposition in technical terms, thereby making it secret; the result was that he was not only defeated, but his loyalty to the US was questioned in a painful and prolonged hearing set in motion by supporters of the H-bomb. Today, more publicly, a number of scientists are fighting a losing battle against "Star Wars" weapons around the question of whether they will work.

Scientists and technologists have a stake in pretending – often to themselves as well as to the public – to be apolitical; they then can hope to be able to decide the course of their own efforts with a minimum of outside interference. But the price is that ultimately they lose any chance to influence the social consequences of that effort – in effect, their own brain children are lost to them. No one can retain control over the consequences of his or her acts forever; but the acts do occur in a social context, with certain values paramount. To maximize one's opportunity to influence outcomes requires attending to likely consequences in advance, and shaping one's actions accordingly.

Both the public interest and the interest of scientists and technologists require that political debates not be disguised as technical ones; for that to happen, the values associated with technological or scientific projects have to be laid out more clearly; those to be affected by a technology must have a role in its formulation; and lastly, technical training must include attention to values and to consequences. These themes will be taken up again in Part Two.

6 Technology in current politics

The right

The Reagan administration took office just as the "high technology revolution" was fully apparent. At this crucial stage, it has inserted conservative values into almost every realm of government policy. These values are far from being completely consistent, but they can be summed up in the idea of shaping the transition so as to strengthen and preserve those sources of power in our society that were being threatened by the egalitarian and environmentalist movements that emerged in the late fifties, augmenting earlier New Deal thrusts in some of the same directions. The administration has attempted to increase the power of corporations and the wealthy by revising taxes, opposing regulations, and cutting back even what Mr Reagan himself terms "the social safety net." It has attempted to reassert US world dominance, at least in the military sphere. It has stood for returning power in families to men, and power in communities to whites.

These policies seem to be attractive to two distinct groups. One is all those who are likely to gain significantly. The other group – largely middle-income workers – are those who had come to resent earlier government programs. Their own social positions inevitably had declined relative to those below them who had benefited from social programs, and they were offended by policies that, however well-intended, had often been poorly thought out and badly executed.

Reagan administration policies in the area of science and technology have been all of a piece with their general views. Democratic and humane values had long been less evident in technology policies than elsewhere in the government; now their influence is still lower. Applied science and development funds have been shifted more decisively towards the Pentagon's control. Alternative technologies that were just beginning to gain a foothold in previous administrations have been substantially cut out of the budgets of the Departments of Agriculture, Energy, and Transportation. These include projects ranging from organic farming, to energy conservation and solar power, to bicycle transportation and even mass transit. Practices supportive of large-scale industry – not necessarily high employment industries, however – returned to unchallenged prominence.

The administration has also been strong in seeking to protect and enlarge the scope of intellectual property laws – for computer "architectures" and software especially; upgrade trade secrecy enforcement; and erect barriers to the transmission of technology to any Soviet allies. For the first time, power over export licenses has been granted to the Defense Department instead of the Commerce Department, and under Caspar Weinberger, the Pentagon has been very broad in its judgments of what technologies have potential military significance. In addition to instruments or machinery, export controls now can also include ideas, and the administration has interpreted the law to give it the right to prevent free exchange of ideas with any foreign national – even within the US – who might then conceivably pass the material on to the Soviet bloc. Conferences can be restricted and scholarly publication subject to advance review; even visits by individuals to college campuses have come under scrutiny. Outrage in the academic community caused the administration to temper these practices, but the degree of restraint it has imposed on itself in this matter turns out to be small.

Given the international character of high technology, efforts to restrict technology transfer are probably as much self-injuring as damaging to the Soviet military. What they really help to effect is the unavailability of technological knowledge to small foreign firms and even to Americans outside existing high technology centers. In addition, they probably slow the development of new technology in places like "Silicon Valley" which have always depended on the relatively free interchange of ideas among engineers working in different firms. But a side effect of that slowdown is again a strengthening of the larger corporations, inside which free discussion can still take place.

In the field of regulation, the administration has not stopped with its well-known reductions of environmental and safety regulations. It has acted with at least as much fervor in deregulating communications technologies when the basis for regulation was some semblance of equality of power or access. The breakup of AT&T, which resulted from an out-of-court settlement between the administration and the company, along with the attempt to strip all local regulations from cable television, the attempt completely to eliminate any public service accountability for radio broadcasters, and the opposition to government-supported, independent public broadcasting all work to increase social dominance by large corporations and the wealthy.

In a similar vein, computer crime legislation favored by the administration and passed at the very end of the Ninety-Eighth Congress makes disclosing data of a nonpersonal, unclassified nature obtained from computer networks, even government computers, a crime. This not only thwarts the new technology's potential to democratize information, but creates a new way to hide materials of public concern from scrutiny. (The administration has also sought to weaken the Freedom of Information Act, to use lie detector tests to prevent leaks – even of unclassified material – from the executive branch, and in general to try to curtail the public's right to know.)

Support for independent work in the social sciences, for

science education, and for values considerations relating to science and technology has never been strong, but what programs there were were eviscerated by the Reagan government as soon as it took office. It is not difficult to see why such programs aroused their dislike. Now, however, perhaps because economic policies of the administration, vague as they are, seem to center on supporting increased technological competition with Japan and Western Europe, concern for improving dismal high school and even college level science teaching has led to reinstatement of some support for science education.

Reading of these initiatives for the first time, a person might understandably conclude that the administration in fact has a bias against science and technology. Support for such a view might come from noting right-wing opposition to the teaching of evolution, which was much in the news in the early days of the administration. Further support might come from the President's reported interest in astrology and Old Testament prophecies. Perhaps further evidence might be seen in the enormous difficulty the administration had in filling the office of President's science advisor – apparently more than ten candidates turned down this previously prestigious post.

But, while the administration may be as inconsistent in this domain as it is around such issues as the national debt, as a whole it seems to be utterly fascinated with science and technology, and eager to exploit this fascination for political purposes. Science and technology were mentioned with enthusiasm not only in the 1984 campaign, but in the President's 1985 "State of the Union" message to Congress. Such fascination is not restricted to the President, who was for many years a spokesman for General Electric. It extends to much of the new, and even not so new, right. Beyond a sheer love of gadgetry as insignia of power and wealth, this fascination amounts to three substantive hopes that the right places in technology.

One hope is that, as in the past, American technological inventiveness will lead to economic preeminence. In the main, this is to stem from new products – for instance, medical devices – that the rest of the world will want to buy. To a slightly lesser extent, high productivity is seen as a means to recapture markets.

A second hope is that the power of organized labor to set wages and control working conditions can be crushed by new labor-saving technology. In a way, these two hopes taken together amount to the hope for a transition to a new, more viable social structure mentioned at the end of Chapter 3. Needless to say, it is assumed that such a hope is consistent with other right-wing values: limited government interference and regulation, no limit to wealth for the wealthy, international dominance for the US, possibly coupled with restored male-dominated family life, school prayers, reduced Social Security and welfare, and no abortion. At least in private, members of the administration accept with equanimity that in the course of this transition, many formerly well-paid blue collar workers will be permanently unemployed, and many other workers will find jobs only at much lower wages.

The most optimistic form these ideas take, represented in the work of Julian Simon and the late Herman Kahn, is that technology can solve all problems: there need be no restraint on population growth or resource use, and of course no need to alter the basic approach of multinational corporations to the environment – or for that matter to alter the basic thrust of the arms race. This universal technological optimism in fact is remarkably similar to that of certain Maoists of the late 1960s and early 1970s. (What is valid in such positions is that projections of ecological doom that ignore the possibility of changed social – including technological – practices are themselves too one-sided.)

The third hope is simply that technological supremacy can restore military supremacy to the US. If we can find the right technology, we will simply not have to deal seriously with the Soviet Union, we can ignore the nuclear freeze campaign, and, of course, defense spending can go on forever. This hope finds its fondest expression in the "Star

Wars" program – which has seemingly won tremendous public support, even though no one appears to be quite clear about what it really is. What makes it so successful is that it is no definite weapon, even no definite concept, merely a research program to see if a means cannot be found to prevent Soviet missiles from striking the US. Whether civilians or only missiles are to be protected is not clear – even though opinion polls indicate that public support is premised on "Star Wars" defending civilians.

As a research program, the expenditures already planned are staggering – for 1986, the administration asked for $3.7 billion, compared with only $1.5 billion for the entire National Science Foundation budget – and that fact probably explains some of the program's support in the scientific and technological communities. Precisely because the program is so loosely defined, the prospect of ever larger expenditures, stretching for years, is great. Ordinarily, a call for a weapons system of this magnitude could be expected to invoke thoughts of an accelerated arms race, but the President has apparently forestalled that with the suggestion that the system could be handed over to the Soviet Union. Quite apart from the fact that the military is showing no signs of being eager to share the technology, should it exist, it is very difficult to see how such a promise could ever be believable.

"Star Wars," whatever it turns out to be, will certainly be a system testable only by full-scale war – i.e., not testable at all. Reliance on it could only come about through faith in computer modelling of what in itself would be an immensely complicated computer system (see Chapter 4). Since the Soviet Union is not located in the same place as the US, and faces a different enemy, it would need different computer programs. There is no reason to suppose that the Soviets would or could ever trust an American modelling of the Soviet version of the program. In effect, there will be nothing the Soviets could have any faith in that could be given to them even if the Pentagon were so generously inclined; the Soviets' only chance of obtaining a system they could rely on would involve racing with us to complete their own "Star Wars" program as fast as we do or faster. Should they fear failure, what they would then do is a very major worry.

In moving towards all these goals, the right as a whole is simply unworried by problems – whether these be social, ecological, or simply logical. They would argue that the US achieved greatness by not worrying and by relying on Adam Smith's "invisible hand" of the market for any necessary correctives. That current optimistic economic projections, and even the optimism of the stock market, are based on continued lavish military spending is a paradox they are happy to ignore. Their current strength results from the ease with which some individuals can now become wealthy (or imagine they can) and therefore feel able to ignore problems dumped into the invisible – but real – hands of the poor, the even more invisible environment, and the human future.

The liberal and neoliberal response

The Democratic coalition reached the zenith of its power during the long post-World War II period of economic growth. During that period, all liberal goals, such as equal opportunity and adequate Social Security, seemed to be contingent on increased productivity, which would lead to higher standards of living and, through increased exports, to more jobs. Military spending would not only assure American access to markets and resources, and protect democracy elsewhere, but would keep the whole system working. Any remaining problems could be solved by government fine-tuning of the economy and social programs. As movements such as civil rights, feminism, environmentalism, and peace led to various fissures in the coalition, the unified world view ceased to make much sense. Looked at today, it seems more like the right's current agenda – except for the commitment to government correction of the market and to social spending for the poor. It is no wonder that Democrats today are confused; simply being a Democrat does not reveal which part of the old view to hold on to.

Given this background, it is not surprising that many Democrats have adopted variants of one or two or even all three of the main right-wing hopes in technology described above. To a greater or lesser extent they have coupled these hopes with a continued concern for some more liberally identified issues. Particularly at the state level, Democrats in office have attempted to improve their own state's competitive position by trying to duplicate the success of California's "Silicon Valley" and Massachusetts' Route 128. At times, in doing so, they have ignored a number of salient facts about the existing high technology centers. These typically include: low wages for manual workers; a number of health and environmental problems that as yet have not been adequately dealt with; dependence on military and space spending as a critical part of economic survival for the industries in these regions as a whole; and the small probability that all states could simultaneously sustain major growth in the high technology sector, and still find markets, much less the needed technical expertise to sustain such rapid expansion. Finally, they completely ignore the question of what the way of life in the post-transition America would be.

Many Democrats concerned about rising unemployment have advocated worker retraining programs that go along with, but are conceptually distinct from, the notion of high tech centers in each state. The idea of retraining workers so that they can find decent jobs is better than ignoring them, but there are several drawbacks. If jobs depend solely on the choices of managers and technologists about efficient ways to work, there is no guarantee that workers who need the jobs will be able to do them, even after retraining. In essence, retraining assumes that workers are a resource for industry; retraining works as a means to assure industry an adequate supply of this resource. The supposition is that a given worker is a blank slate who can be retrained to any kind of task. For instance, a steelworker can be retrained to be a computer programmer or a chemical analyst. Whether the worker wants to do this kind of work, or whether most steelworkers can make such adjustments – not to mention whether there will really be a job after retraining – is almost never considered.

If workers are regarded first and foremost as people, then instead of shaping workers to fit supposedly preexistent slots, a more rational aim would be to fit jobs to the abilities and desires of workers. Job redesign would take priority over worker retraining. Such redesign, in turn, would entail shaping technology to jobs and not the other way round.

In short, Democratic hopes for a good society riding on adopting the "Silicon Valley" model appear to be based on poor observation and invalid generalizing from what will very likely remain quite limited phenomena. They have seen regions of bustling economies, assumed that a bustling economy would automatically solve all social problems – since all problems supposedly stem from a shortage of jobs – and then assumed that the same solution would work for everyone everywhere. Unfortunately, once a commitment has been made to establish high technology industries, attempting to find such industries can lead to ignoring the problems associated with them, and can further increase economic dependence on military spending to subsidize high technology. A vicious circle can then be set in motion. In the course of this process, social goals can fall further and further by the wayside. Another path is needed.

Quite a few Democrats believe they have found that other path in the idea of industrial policy. Industrial policy represents a substantial improvement on the damaging model of each state competing to lift its economy with the same industries. The idea of industrial policy is that somehow the national government should decide on the mix of industries to encourage in the country as a whole – taking into account the needs of various groups and the availability of resources – and then promote policies that will further these goals. These policies do not include the rigidities of Soviet-style planning: for example, quotas would not be set, and market mechanisms would continue to determine prices and amounts of goods produced. Instead, the components of government industrial policy involve such features as loans to new industries, government-sponsored training programs, and government-sponsored research into carefully chosen technical areas that will be needed to create new industries.

The basic premise of industrial policy is that the US position in the world economy is being eroded by the rapid development of other countries, starting with Japan and Western Europe, which compete on the basis of more modern factories and better cooperation among corporations, government, and labor. Additionally, many new industrializing countries, such as Singapore and South Korea, have significantly lower labor costs. The main conclusion reached is that further unemployment and economic decline will result from the failure of the United States to modernize its industries on some rational basis. Modernization in turn has two elements: increasing productivity and replacing declining industries with newer, generally high technology industries. Advocates differ on exactly what form rationalization should take. Robert Reich has argued in favor of widespread democratic involvement in decision making, but has not made clear exactly how that would work. The AFL-CIO and centrists such as Felix Rohatyn have proposed a less far-ranging but more centralized industrial policy, the main feature of which would be a tripartite labor-industry-government bank able to make loans to support targeted industries in specific locations.

There is much that is commendable in the notion of industrial policy; certainly, the values that underlie it are more humane than those of the Reagan administration. However, there are also serious problems that at the very least require a major rethinking and expansion of the notion. First – to repeat a major theme of this book – the idea of new industries to replace declining ones implies a specific vision of the future. Without such a vision it will be impossible to choose what to produce; if the vision and the values underlying it are not spelled out, there is no reason to think the overall results will be desirable. The new products and new production processes may work against equality, democracy, and other central values. There is nothing in the institutions advocates of industrial policy propose that will deal with this concern.

Second, increasing competitiveness through higher productivity and new industries may worsen rather than improve both American and world unemployment problems. If new industries and new production methods require new skills, then American workers are not necessarily in a better position to learn these skills than workers available at far lower wages elsewhere. Meanwhile, existing industries' present level of productivity is enough to undercut the livelihoods of millions of peasants and hand-workers in other countries; any substantial increase in that productivity would make the damage that much worse.

Third, industrial policy involves adopting a model that worked in other countries under very different circumstances: they were attempting to catch up and not, until very recently, to lead in the choices of new industries. They already had a model of what industries to support in order to obtain a desired result – similarity to the US. Since the US had an extensive democratic and egalitarian tradition, that went along – to an extent. But there is no guarantee that industrial policies now underway in France and Japan will lead to more changes in the same direction. In fact, both Japan and France have long histories of a much greater degree of economic centralization than the United States. This makes policies possible there that would hardly promote democracy or equality here.

Another distinction between the countries that have previously adopted industrial policies and the US relates to employment problems. The two leaders in industrial policy both maintained large agricultural sectors as a matter of policy, conditioned by the resistance of farmers to giving up their ways of life. Both countries also relied on pools of surplus labor that did not show up in their unemployment statistics; foreign workers in France, and workers over the retirement age of 55, as well as women, in Japan. Geographic barriers make it easy for both countries to regulate immigration (virtually nonexistent in Japan), which leaves them with aging workforces. Finally, France, like other European countries, is increasingly specializing in the production of capital goods, the export of which will only speed world-wide competition. Were the US to attempt to compete for the same markets, that would only intensify competition, and quite probably unemployment, still further.

A fourth difficulty in industrial policy formulations emerges from that French policy as well. The underlying assumption is that Third World countries will replace the advanced countries as sources of basic consumer goods, but that these countries will provide a market for high technology goods, including capital goods from the advanced countries. In order that the advanced countries can maintain an overall positive trade balance, the Third World countries would either have to sell at reduced prices, or they would have to buy huge quantities of high tech goods and services. In the latter case, they would face substantial indebtedness perpetually. In other words, the backbone of current industrial policy notions is continuing inequality among nations, even while development occurs. Not only does this seem morally dubious, but the built-in instability would pose an ever greater threat.

The inequality underlying industrial policy is not necessarily for the Third World only. Without careful and explicit efforts to the contrary, an industrial policy could easily end up providing good and well-paying jobs especially suited to the white males who have recently lost ground; the institutionalization of practices implied by the policy could help maintain the inequality of the past still farther into the future. Even those who have the "good" jobs will only be well served if these jobs not only pay well, but are worthwhile in themselves, available without having to abandon family and community ties, and do not require disorienting and ego-destructive forms of "retraining." None of these conditions is a necessary consequence of most of the industrial policies now being considered.

The realities of the US make it necessary for there to be a substantial political struggle to bring into existence any but the most procorporate of industrial policies (which we really already have). For such a struggle to succeed and be worthwhile, some of the problems mentioned above ought to be addressed first. This will require a much more comprehensive set of policies.

Few Democrats have gone beyond employment issues when it comes to formulating positions relative to new technology, except in the area of defense. Here, Gary Hart has been most notable in arguing that "more isn't better; less isn't better; better is better." What this presumably translates into is the view, which a few neoliberals seem to have arrived at by suddenly rediscovering the great military theorists of the past, that rather than be mesmerized by the gimmickry of high tech weapons, we should ask whether they would really improve American capacity to fight the wars we would want to fight. That there necessarily would be such wars remains an unquestioned assumption in this formulation, but that is the issue that should precede discussions about the right weapons to purchase. Also, military enthusiasm for high tech weapons, as mentioned in Chapter 3, is a partially rational response to the reality that Americans quickly lose enthusiasm for dying – or watching dying – in protracted guerrilla wars, the only kind, other than a nuclear war, that we are at all likely to get into now, and a circumstance that the newly resurrected tacticians never encountered in their times. The reality would only change if life in the US became sharply worse for prospective soldiers, which is something we should concentrate on preventing. Hart is wrong; less is better.

These and a few other Democratic proposals, where they differ from right-wing proposals, are usually superior in that they recognize the importance of equality, social caring, and democracy itself. They fail only in that, as a whole, they have been insufficiently thought through. They continue to assume that growth based on international military and commercial competitiveness is both feasible and desirable. They accept too readily the primacy of management and industry views in deciding the best direction for the country. They do not look sufficiently at life during or after the technological and social transition that they join with the right in postulating. The remainder of this book is intended to sketch out – in preliminary form – the shape of a possible alternative.

PART TWO

Reinventing technology

7 Introduction to Part Two: The basis for an alternative policy

We have come full circle. Starting with the observation that politicians of both parties are emphasizing science and technology in their oratory, the argument has proceeded through an analysis of the nature of technology, examined its major impacts, turned to the special case of high technology, then to the role of government in fostering the sort of technology we have, and at last arrived at an account of the actual political programs that relate to technology. The conclusion is that present and prospective policies are distorted. They are not moving us towards a more just world, and may be seriously endangering even the degrees of equality, democracy, and world peace we currently enjoy.

Technology policy, then, ought to be redirected. But in what way? There seem to be two basic possibilities. One – which finds occasional support on the left – is to say, in effect, "if current technology is pernicious, why not end any technological development altogether? After all, the human race has survived thus far without all the technologies that haven't yet been developed. Simply freeze technology in its present form, to limit the damage." That position has some merit; certainly there are areas, such as nuclear missile development, where a freeze would be salutory – in this case a vital step in ending the arms race. But for technology policy as a whole, the prescription, if followed, would probably increase rather than decrease human misery.

Assume, for the sake of argument, that a technological

freeze were somehow adopted politically. Our economic system, based as it is on continuous growth, would immediately begin to collapse, as demand for new kinds of products and services and for new production facilities began to shrink. A freeze would only be tenable if a new social order were simultaneously to be created. The new technologically static order might turn out to be either a form of feudalism or a form of socialism or something different from either. But because the new order – whatever it is – would have to make do with existing technologies, possible social relations would be limited. Technology, of course, involves both processes and instruments, or to put it another way, not only things but social practices. Restricting ourselves to currently existing technologies amounts to retaining some of the worst features of our society, while pretending to try to live differently. Abandoning technology altogether, in a return to the pretechnological era, would be even worse: there are about four billion more people alive now than at the beginning of the technological era; many of them could be expected to die if agriculture, medicine, and the distribution system suddenly were to revert to earlier forms.

Any realistic social order will in practice have to find ways to redirect technology if it is going to avoid disaster or being stuck forever with some of the worst aspects of the current order. We are left with the second possible policy option: a full-scale program for redirecting technology according to the values and social visions we would want a new order to exemplify. (This would be true even for a new order that differed in only a few respects from the present one.) That program could conceivably involve, in part, going back to pretechnological practices and modifying or developing them to be sustainable in different conditions. (That is the program of the "appropriate technology" movement.) But if past social orders had flaws – including isolation, inequality, and lack of effective democracy – which we would not like to perpetuate, then we are going to have to find ways to utilize and reshape much of current technology as well.

We need new institutions, new programs, new technologies. In short, we need a way of reinventing technology as a social system so that it can better meet our needs. Part Two fleshes out what that would entail, through the device of a model policy program. The aim is less to get this precise program drafted into law than to show that a real alternative to the present system is possible, and – at least for those who agree with what they have read so far – clearly preferable. While rapid legislative adoption of a program like this is highly improbable in the current political climate, without clear alternatives as possible rallying points for opposition that climate itself is unlikely ever to change.

Details of the program are presented in Chapters 8 to 13. The remainder of this chapter is a summary of the values and assumptions which underlie the program that follows. In part, it will be a restatement of some of the conclusions reached earlier, and in part a clarification of what is meant in this work by "democratic values."

Democracy and technology

Democracy is a centrally important value, and has long been accepted as such in America. It is a highly practical ideal: it offers a means for social disagreement to come to the surface before society disintegrates; it offers a chance for social relationships to be continuously reconstructed to accord with currently felt needs; it is opposed to the concentration of power in a narrow class or a few individuals, and limits exploitation and oppression. It is an ideal that incorporates within it other key values such as equality, community, free expression, cultural diversity, and peace, since without these democracy would be meaningless or impossible. But, like other ideals, democracy can never be completely realized. Furthermore, at a given historical moment, the question of whether or not we have democracy depends on a comparison to what would be possible: would there be a conceivable way to distribute power more widely and more evenly, and to include in its exercise more of the issues that really matter? As times change, democracy, to survive, has to be continuously reformulated.

The democratic ideal entails that each citizen should have an equal chance to be heard on issues of importance that confront the political unit in question. This simple formulation implies all the other key values listed above. Each citizen's opinions are to count. For that, everyone must have a chance to form an opinion, which implies having an opportunity to know and understand whatever issues he or she has concluded might be important. This requires that leisure, information, the privacy to contemplate, and a venue to engage in discussion or debate, all be available to everyone equally. Democracy loses its meaning if a society loses its commitment to equality, if information is concentrated in too few hands, if issues are decided in such technical form that most people are unable (or imagine themselves unable) to understand what is being argued, if certain areas are declared out of bounds for decision.

Democracy does not imply homogeneity; different viewpoints, even different ways of seeing the world, are not only acceptable, but necessary for democracy to have any meaning. However, it does imply that cultural differences must not be so great as to allow no intercommunication at all. Democracy can work only if each person is in a position to understand the world as involving a group of interrelated and interdependent communities, and through that to grasp her or his relationship to ever wider circles of others.

That relationship cannot be overly antagonistic. War and democracy are inconsistent; not only does warfare require hierarchical decision-making, but in war a society loses to its enemy the power to decide issues deeply affecting its own future, and this can be true even for the ultimate victor. In a global economy, democracy cannot stop at national borders; without shared policies, democracy in each country can be overruled by the forces of international competition.

How are these ideals affected by current realities, particularly those that relate to technology? Since supporting democracy has not been a high priority in deciding the direction of technological development, it is not surprising that existing technologies pose many obstacles. First, it should be obvious by now that the detailed direction of technology is so decisive a force in people's lives that for democracy to continue to be meaningful, it must include ways to affect those detailed decisions.

Second, new technologies have continued to help undercut the importance of local communities. Increasingly, people's sense of community is not determined by geography, as it once was; in the United States at least, few can be expected to live out their lives near their birthplaces. That implies that choices made at a local or state government level are less significant for their citizens than they once were. For example, despite current concerns about education, no state or local government has much reason to believe that the children it is responsible for educating will continue to live there, especially if their education provides them other options. So the impetus for good education must come from the nation as a whole. Similarly, people have to look to the whole society for a wide variety of services and assurances – such as pension rights or air traffic control – that simply would not mean much if available only locally. Since a technological innovation spreads rapidly from the particular place it was made, control of technological innovation especially requires action at the broadest level, i.e., at the level of the federal government, or through international bodies.

The absence of strong ties to local communities also means that a system of democractic government based on representation of particular geographic districts is in reality less democratic than it once was. In effect, representatives do not have clear constituencies any more. That is one of the main reasons "interest group" politics have become so dominant. The only problem with this development is that not all potential interest groups have equal opportunity to form themselves or to exert influence.

Third, since local communities are weaker, the problem

of each person having a recognized and established place in the larger society is far more challenging than ever before. It is difficult for democracy to function if a large proportion of the effective community is invisible to other members of the same community. Finding new means for people to integrate themselves into some sort of community, local or not, becomes an essential task.

Fourth, life lived in a global community has become far more complex. Each technological innovation requires new decisions. This level of complexity means that the democratic forms we have are no longer equal to the task of allowing each person to have a significant voice in matters of concern. If decisions are to be made on a federal level, we can hardly suppose that 535 Senators and Representatives can possibly cover all the necessary topics; certainly, Congress as a whole does not have the time to deliberate everything. How to construct new forms of democracy that would allow some influence over all the decisions that affect us is one of the daunting challenges of our time.

Fifth, as Chapter 4 indicated, we seem to be moving rapidly in a direction of greater inequality of wealth and access to information. For democracy to remain alive, that has to be changed. A society that respects all its members cannot accept trends that increase relative, and therefore eventually absolute, powerlessness and poverty.

Sixth, the continuing international arms race is increasingly at odds with democracy. Decisions about nuclear annihilation that must be taken – possibly by obscure military officers – within a half-hour warning time obviously bear no semblance to democratic choice. Demands by military institutions for funding essentially secret new projects designed to offset some equally secret, presumed advantages of an opponent limit and contract the options for democratic choice.

Finally, technology itself intrinsically includes an anti-democratic element: decisions affecting large areas of people's lives – most especially their work – are made without their direct involvement. As technology continues to grow in importance, finding ways to change it to what amounts to an inherently more democratic form of innovation is a more and more pressing necessity. While such a transformation would alter relations between technologists and the rest of society, it should also open new opportunities for them to employ their skills in worthy ways.

Democratizing technology

The last section described some of the ways in which technological developments have limited democracy. At their core is the lack of influence that democratic values and interests have had in the direction of technology in the first place. If new policies are to help lead to different results, they should be constructed democratically. Groups who had little opportunity to influence technology before ought to get more power. However, there doesn't seem to be any possibility that such groups could quickly arrive at true equality of power over technology, and if a move towards democratization is eventually to succeed, this difficulty must not be disguised or glossed over.

Existing "special interest groups" who do have power, such as the National Association of Broadcasters, are able to lobby not only Congress but agencies that have specific authority over technologies of interest to that group itself, such as the Federal Communications Commission. The broadcasters' association has a high degree of cohesion, a clear sense of its common interests, excellent information regarding the timing of decisions potentially relevant to it, the resources of time and communicative capacity to work towards a consensus, and virtually unlimited amounts of technical expertise at its disposal. At the other extreme, groups of the poor – say, welfare mothers – who might have an equal stake in some technological choice, if only they knew about it, have none of these advantages, and cannot obtain them simply by legislated modification of institutions, no matter how vast. For example, technical expertise within the group would not suddenly grow, and in fact could never be as widely available to members of the group as it could to a more narrowly focussed interest group.

If the relatively powerless cannot obtain equal power over technology, a program to change the existing direction of technology might simply end up shifting power from one relatively small group in society to some other group. To prevent that, a conscious effort to help open up the process of technological decision-making would be needed. To that end, the program described in the next six chapters incorporates the following measures:

1 Specific technological programs would have well defined, understandable goals – e.g., technology for worker self-management or for heightening democratic involvement in the political process.

2 For each program there would be a corresponding agency set up specifically to foster it (and named accordingly). That should aid in assuring public accountability and also ease access to the work of the program. It should help as well in at least partially assuring staff loyalty to the program's aims.

3 All the programs are intended in one way or another to help promote democracy, equality, and the other key values mentioned in the previous section. Among them would be programs for providing communications facilities suited for improving the capacities of groups to organize themselves and facilitating access to a wide variety of information, presented in a wide enough range of forms to be intelligible to people of diverse ethnic and educational backgrounds.

4 Each agency would have a board chosen in the most democratic manner possible to represent the currently disempowered groups likely to be affected by the work of that agency. A fuller description of how this might be organized will be found in Chapter 13.

5 One agency would have the specific responsibility to develop additional measures to promote direct contact

between any of the agencies and the members of groups not sufficiently organized to communicate as a body.

6 There would be a number of legal requirements, such as social impact statements for new inventions, and guaranteed rights to a voice in technological decisions (e.g., for workers in factories), that should help spread knowledge of, and power over, private sector innovation.

7 Since the purpose of this program is not to dictate the nature of future society, and also since particular agencies might differ sharply in their rate of progress towards their stated goals, a considerable amount of redundancy and overlap between programs is intended. This would help assure a diversity of technological options. (While such duplication may seem unusual for revenue-starved social programs, it is customary at present for military programs.)

8 The final guarantee would be public awareness. If these measures, taken together, were to succeed, then equality of power, however measured, would increase noticeably. Failure should also be evident and would be difficult to disguise after a while. If any genuine public commitment to strengthening democracy and increasing equality survived, it would then lead to further efforts.

Bureaucracy and policy

The policies offered in Part Two assume a substantial bureaucracy. A well-known characteristic of bureaucracies is that they tend to grow – often at the expense of their intended purposes. But they also help assure that the values they serve continue to get some hearing. Many of their members do come to believe in what they are working for. And no policy can be carried out without some bureau dedicated to seeing that it is. The answer to problems of bureaucratic inertia can only be some system of competition among different offices, accountability to the people in whose interests one is ultimately working, a sense of mission, and a system of checks and balances between various parts of the bureaucracy. To the greatest extent possible, all their functions and assignments, as well as their actual accomplishments, should be intelligible and accessible to the wider public and to the communities they particularly serve. The policies offered here attempt to meet all these criteria. Bureaucracies that serve good and worthwhile ends and have feasible assignments are better than those that serve more dubious ends; they may attract more dedicated workers, and may also provide an environment in which it is possible for workers to be rewarded by seeing their efforts come to fruition. These bureaucracies may be less frustrating and enervating places to work in, and may be less likely to fall prey to inertia, than many existing ones.

Technology and equality

There is no aspect of the myth of America so powerful as that it is a land of opportunity. As in many myths, there is more than a grain of truth to it. As a country that had no preexisting class structure of any rigidity, it has been unusually open to those with sufficient ambition, industriousness, wit, and luck to rise fairly high, although how much of each of these was needed still depended on race, sex, and other elements of a person's background. As was pointed out in Chapter 4, much of the drive towards lowering barriers that occurred in the past century or so could probably be linked to the consequences of the technology of mass production. It is perhaps not surprising that even now technology is viewed as the great elevator of fortunes, part of the rising tide that will lift all boats. The trouble is, of course, that it could just as well help sink many boats.

There is no doubt that, in a period of turmoil and rapid change, the smart and ambitious have a chance to rise relative to others. One question we must ask ourselves is whether these are the only attributes that should be rewarded, or, to put it the other way round, whether the absence of these attributes deserves punishment. As was discussed in Chapter 3, relative poverty becomes absolute, and that is especially true when the barriers to falling or rising are removed.

What do we really mean by equality? The usual phrase in which this word occurs is "equality of opportunity," but even that is a very loose term. When is this supposed to exist? At birth? Until graduation from high school? Throughout life? These are very different concepts, but if we are truly to have a society free of discrimination on the basis of gender, race, physical disability, or age, then we had better mean that equality of opportunity continues throughout life. (Even that is hardly fair, when for reasons that certainly have much to do with government policies life expectancies are not equal for the rich and poor or for whites and blacks. As health is not the topic of this volume, that point will not be pursued here.)

Exact equality of opportunity throughout life would imply that a person not be punished for past mistakes, bad luck, lack of motivation, or the consequences of prior discrimination. Likewise, it would mean that a person not gain (relative to others) for past good luck, wise choices, high motivation, or benefitting from discrimination. These conditions could never be completely fulfilled, for people are in many ways themselves the result of their past, and that can hardly be completely overcome or lost. So any reasonable equality of opportunity would have to mean that our society simply would not offer too substantial external punishments or rewards for what a person had made of him or herself in the past. In other words, if we seriously mean equality of opportunity, then we would have to mean rough equality of treatment for every person.

Equality of treatment means, approximately, that each person would have the same chances for satisfaction, insofar as that is within society's power. Equality does not mean identity. People can be satisfied only if the differences between them are taken into account. So equality does not mean uniform goods for each person. It could begin to be approached only if a very wide class of people's needs could be adequately met. That brings the argument to the issue of scarcity, which will connect back to technology. There are obviously certain kinds of satisfactions that everyone cannot share in. For instance, not everyone can have servants, at least if the servants are people. Since land is limited, not everyone can have a thousand acre estate or farm. Not everyone can be famous. If these are the sort of satisfactions that we insist on ranking highly, then inequality is inevitable.

But there are things for which equality, in a meaningful degree, is possible. Right now, everyone can have white bread. Everyone can watch television. For many, these are dubious satisfactions at best; for no one are they enough. Where technology comes in is in helping increase the number and variety of things that everyone can have to the point that it might be palatable to accept equality of satisfaction as a realistic possibility, or at least to move closer to it.

An obvious problem is limited resources and environmental consequences; but even when environmental preservation also is included as a requirement, there still would be possibilities for technologies to increase the variety of available satisfactions.

The role of competition

Careful readers will have noticed two apparently contradictory views of competition in Part One. First, technological development for the purpose of becoming more competitive (e.g., as a nation) was characterized as counterproductive, since it results in increasing the level of competition in the world, which ultimately increases world unemployment. Similarly, military competitiveness only encourages arms races, in the end lessening everyone's security. Second, it was held to be inappropriate, for example, that some large corporations use technological supremacy and intellectual property laws to ward off possible competition. A variety of different technologies for achieving worthwhile ends was held to be good, even if that variety were developed by companies each striving thereby to increase its own competitiveness.

Is there a way to reconcile these two views? The first looks at the effect of competition in terms of the long-term interests of one of the parties. Contrary to commonly held ideas, it asserts that in those terms competition is destructive. In fact, this is hardly news. Firms and nations compete precisely to achieve the elimination of competition which they recognize as an obstacle to their long-term survival. The only problem is that their short-term actions are clearly gambles that rarely succeed. Getting free of competition means that most firms, or most nations, eventually lose out. In a full-scale struggle, both sides can lose.

Supporters of a completely free market and the value of unbridled competition, without government interference, emphasize and even glory in the freedom not only to succeed but to fail. It is failure that in their view eliminates the bad ideas, the unsound practices, waste, or inefficiency; failure prevents ever more resources from being pumped down what they perceive as ratholes. There is something to be said for this view. It is perverse for society to continue in a bad direction simply because of a heavy past investment. But failure eliminates more than unsound practices. It undercuts a chance for a decent life for workers in failed firms and for citizens of failed nations, regardless of whether they were at fault. It destroys firms, enterprises, and nations that in many respects might be better than their surviving competition.

When rises in productivity in highly competitive economic sectors lead to overproduction, even quite well run firms, farms, or individual efforts can go under. The threat of failure is often paralyzing; equally often it encourages desperate measures – cheating, shoddy standards, unsafe conditions, sabotage of competitors, and so on. As this process continues, failure eventually does lead to the end of competition, leaving instead a few large firms to dominate a market for a considerable period.

The major benefit of competition is to those outside the process. When competition is not so severe as to lead to shoddy or sharp practices, it leads to a meaningful degree of choice for the larger public. In effect, competition in some fields means that whatever public there is for that activity cannot be dictated to by one power center. So, ultimately it is the democratic aspect of competition which is most positive. Permitting real, meaningful choices is a good thing. Price competition or, equivalently, denying the power of some cartel to determine prices and therefore to determine who may have what, is also good as long as it does not result in enough failures to lead to monopoly, ending the benefit for the outside public. If low prices are achieved by low wages, the wider public, which of course includes workers, is no longer really outside the process, so that the apparent benefits are not real.

Competition in less extreme forms does seem to have something to offer, aside from simply being fun, even for people on the inside. When the stakes are not too high, the urge to prosper, or outdo an opponent, often increases inventiveness, care, and zeal in serving the larger community. Worthwhile innovations are unlikely to be adopted by monopolies that have nothing to gain. The same probably holds for people. Human identity always depends in some measure on visible signs of having attained the notice of others. If what I do has no influence on others' reactions, then it may well cease to have meaning for me too. A person who comes to feel that his or her own position in society will remain the same no matter what he or she does beyond some irreducible minimum may come to feel utterly alienated and sink to that minimum level. (This argument might be construed as a reason to abolish income support such as the welfare system. Care must be taken in distinguishing between this and other sources of apparent lack of motivation, such as powerlessness, extreme poverty, or more tasks or responsibilities than one can handle. For most people currently on "welfare," these other sources play a much bigger alienating role than does too secure an income.)

A humane society, then, would not abolish competition, but would increase its scope while limiting its degree. It would aim for almost everyone having some chance to shine, and for failure to threaten no one's survival, dignity, or future opportunities. As was argued in Chapter 3, relative differences in wealth or power eventually mean that those on the bottom also suffer in absolute terms. Hence, for the consequences of failure to be limited, the rewards for success must be limited too. Dizzying success itself poses a threat of subsequent failure, and the lure of enormous success, like the fear of overwhelming failure, can lead to practices counterproductive for society as a whole.

To be humane, a society would have to emphasize rewards other than standard of living or power as incentives to excel. These could include the pleasures of creativity, satisfaction at a job well done, public recognition, the intrinsic interest of the work itself, the pleasure of learning new skills, or of coming to understand the world better.

For all this to be possible requires realigned social priorities; but technology can help make such a world feasible. For example, communication systems permitting cooperation and the rapid spread of ideas and dialogue would help. So would capacities for meeting needs on a small scale; this would permit the formation of communities – not necessarily localized – whose members could have primary responsibility for meeting each other's needs. It would have to be possible for each person to find a community in which to participate, or failing that, to be clearly included in the larger society. Communities would also have to have means of cooperating, sharing information, and aiding each other in emergencies. The right of each person to leave a community would have to be assured by the larger society.

A community-based economic system would help assure each person's control over her or his own life, because each community's economic well-being would depend only to a small extent on forces beyond its control – whether large corporations, government allocations, or the rise and fall of the world market in some specific commodity.

This economic model is also offered in partial answer to the question of whether it is possible for each person to live a good, fully human life. Humans have needs of all sorts that can never be fully catalogued because they are open-ended; that is, people create new needs for themselves all the time. At any given point each society offers its members a certain number of pathways towards the satisfaction of various needs. (In our society the pathways currently include jobs, social contacts, self-help, purchases, education, inherited wealth, welfare, chance meetings, etc.) But these pathways do not generally provide for everybody. Some pathways are full (e.g., a limited number of good jobs); others are not widely known, are relatively inaccessible, or are difficult to navigate. Increasing the number of accessible, intelligible pathways to satisfaction of the full gamut of human needs is essential for an egalitarian, democratic society. Technology could help, and developing the variety of technologies which could enable such pathways to increase faster than the number of people or than unmet needs should be the central goal of an alternative technology policy. Needless to say, that would not include satisfying needs for power over others, or for destruction.

The global setting

Policies offered here are specifically intended for America. To the degree that we have a definite culture and values we wish to preserve and extend, it is appropriate that we concentrate on technologies that would serve those values. But there is no escaping that we live in a finite world. Any workable technology policy has to take into account the rest of the world as well. For the other advanced countries, this can take the form of agreements to place cooperation ahead of competition and to work jointly towards a goal of world-wide equality.

The people of the less developed countries have been trapped in what might be called the tragedy of development. Regardless of what they might freely choose, the only way these countries can avoid further encroachments of powerful cultures that have embraced technology is to try to develop equality in power, which involves at least partially embracing technology themselves. Further, the embrace of development is often, and unsurprisingly, based on ideas of speeding growth in such a way that either dependence on developed countries is increased, or portions of the developing societies are made to suffer terribly, or both.

If we decide to pursue policies of community-based econ-omics domestically, we nonetheless cannot justly abandon these countries to the misery that would now likely be their overall fate without our help; after all, we have benefited from their resources and their efforts. But we must make sure that our help is not a disguise for our rule. We can probably help best by developing technologies that Third World countries could use to maintain their existing cultures without destructive modification, but with a rise in internal equality and basic standard of living. That essentially means helping develop technologies that can enrich life for inhabi-tants of villages, as well as for those now crowded into urban areas. It also can mean offering them unlimited and free use of as much of our technological knowledge, infor-mation, and designs as possible. Villagers could all probably benefit from decentralized communications systems by which they could contact either villagers in other places or distant experts to get help in arriving at socially and environmentally sound solutions to economic problems.

Were the advanced countries to agree now to act together, putting aside emphasis on military confrontation and exag-gerated economic competition, they could almost certainly by early next century arrive at workable means to bring the Third World up to advanced country levels of satisfaction, without forcing cultural homogeneity. That would be a truly worthy national goal – a goal that could make sense in terms of a program such as the one set forth in the next few chapters.

8 A model "Social Goals-Directed Technology and Science Program"

Social ills demand not only attention, but inventiveness. Just as one source of these ills is our current technology, some of the possible cures will also be technological. This and the next five chapters describe the range of government institutional and policy changes that could permit a more democratic and humane technology to emerge.

To describe some of the changes that are possible, these chapters are shaped very loosely as model legislation, either in terms of new government agencies of the sort that usually require legislative mandates, or as laws shaping rights and possibilities for private activities and contractual relationships. This form seems appropriately suggestive, since legislation implies more permanent change, with greater public participation in its formulation, than would either executive orders or simple policy decisions. Adherence to the form is "loose," skipping awkward legalisms and gratuitous details, and adding comments freely.

The order of the presentation follows a certain logic. This chapter concerns policies in which the initiative for developing new technologies and the necessary supporting research comes directly from specific government agencies.

At times the government goes beyond sponsoring development, seeing to the actual operation of the technologies in the country as a whole (e.g., air traffic control technologies or rural electrification). That sort of activity is the subject of Chapter 9.

Chapter 10 takes up the government's role in creating an atmosphere conducive to certain kinds of private sector innovation; the focus is on intellectual property laws. Government must also deal with the consequences of innovation of whatever source, through regulation, compensation, etc. Changes needed in such policies are found in Chapter 11.

The program sketched out in these four chapters might be sufficient, were we alone in the world. In reality, the program must have international dimensions, and possible US government initiatives to promote cooperation and reduce economic and military confrontation are the subject of Chapter 12.

Chapter 13 is intended to clarify some of the workings of the proposed program as a whole.

As explained in Chapter 1, the tasks of detailing technological aspects of compatible environmental, energy, agricultural, defense, and transportation policies are left, with incidental exceptions, for companion volumes in this series. However, many broad principles included here would have obvious significance for the other policy areas.

* * * * *

The premise of the Social Goals-Directed Technology and Science Program is that there are many kinds of possible socially valuable technologies which only the government has the capacity to develop. Not attempting to do so amounts to an injustice; a failure of the principle of "government for all the people." This program concerns positive steps for redirecting government-sponsored research and development; along with them must come a decrease in misdirected technology, as in new weapons.

The key innovation in this Program would be the creation of a number of agencies, each dedicated to coordinating the development of technologies intended to support a definite social goal – either bettering society from the standpoint of a certain value, or working in the interests of a particular social group. Each agency's main tasks would be to define

and then to coordinate the kinds of technological developments that might best help promote its goal. (The next,
and major, section of this chapter will describe these goal-
directed agencies in more detail.) Often, several different
goals would require variations on one particular technology
– say, a means of permitting people more flexible control
over machinery they work with directly; in such cases an
additional, more sharply focussed agency or office could be
set up to develop that technology in the directions indicated.

Since technological possibilities are influenced by available scientific knowledge, both applied and basic research
would also have to be redirected appropriately. That implies
further new agencies to sponsor the mix of research appropriate to the social goals. It is true that the more basic – or
"pure" – the science, the more difficult to gauge how the
results might eventually be of use. But, as there are always
more lines of possible research than can be pursued, choices
are in fact made, and they are often made with applications
in mind. Social goals-directed technology would need a new
science base, and like other parts of the program it can only
evolve if allowed the same imaginative scope that presently
favored technological projects have been able to offer.

All this should involve newly formed institutions as much
as possible. Simply taking over an existing project, bureau,
or even a laboratory – from the military, for example –
would imply accepting many of the values structured into
the organization by its prior sponsors. Starting new institutions also would ensure that social goal-directed projects
do not simply become means to keep programs like weapons
research alive through lean times.

Many of the proposals in this program may seem highly
experimental and speculative. But contemporary weapons
or space satellites would have seemed equally unlikely had
not our society devoted enormous resources and much time
to their development. Technology for social needs must
also have ample sustenance to be more than a gimmick.
Likewise, just as in other areas of technology, some projects
will surely fail; that must be accepted. Over time, commit-

ment to the new goals would lead to changes deep within the structures of science and technology; these changes would then probably permit new applications to sustain democratic values in ways that cannot now be foreseen.

The agencies suggested here would view a number of common problems from different angles, allowing a healthy degree of redundancy. Not all of the resulting technologies would necessarily end up being put into effect. The program implies a general direction of social change while permitting flexible future choices. Flexibility would be limited if too narrow a range of values were incorporated into technologies, since that would make sustaining any other values relatively more difficult.

Encouraging increased democratic control is a paramount goal of this program; so it is important to find means of keeping the proposed agencies accountable to their intended beneficiaries. This aim requires experiment; there is no established method. A special office could be responsible for spurring and evaluating attempts. (Chapter 7 indicated a few options.) The technologies to be developed would offer new possibilities for achieving heightened democracy, so they certainly should go forward while experiments in new forms of control by constituencies go on.

In summary, the resultant structure would include the following: the social goal-directed technology agencies; the more focussed agencies for developing specific technologies; an organization that would coordinate or sponsor more general technological development and scientific research to support the specific work of the project agencies; and offices devoted to such tasks as identifying possible additional social goals, assuring appropriate citizen participation, coordinating goals and values between the different agencies, and assuring the even development of technologies with respect to their overall social impact. (While this list of offices and agencies might seem unworkably complex, in fact it is no more complicated than the administrative structures that currently characterize large technological undertakings, whether corporate or governmental. The details of administration are, of course, different from those of existing programs, as they would have to be to serve the new constituencies, goals, and values.)

Along with development must come other measures. It is pointless for new technologies simply to exist; their intended beneficiaries need to know of them, and must have the power to choose or reject them. That requires legal guarantees of new rights. Chapter 11 presents details.

Even put into use, redirected technology alone cannot bring about a humane and livable world. Complementary social programs need new vigor too. Some of what that might entail will appear or be implied in the specific proposals of this chapter. They would include such measures as broadening and renewal of affirmative action, a transformation of welfare into some form of guaranteed income, increased opportunities for political participation in issues of all sorts. Important as these steps would be, however, they should not be taken to be the comprehensive set of social programs – beyond the scope of this book – that in fact are needed. The movement for improved social programs of all kinds is now mostly dormant; visible support for the goals in this Program could only help restimulate interest in all the rest.

Agencies Relating to Work

Work is the area of life in which most people are forced to come into contact with technologies over which they have little control. While a small and inadequate proportion of federal funds are devoted to occupational safety and health, almost nothing is spent on other areas relating to the nature of work and employment – other than efforts to increase work productivity.

The Office for Improved Work Quality

Machinery is often designed with the ideas that efficient management requires control and initiative to be taken away from the ordinary worker and vested in technical supervisors. This practice demeans workers, while often actually depressing overall efficiency. Such design strategies are especially prevalent in the new generation of automated equipment, where programming and software could easily be under the control of rank-and-file workers but often is not.

Regardless of the preferences of individual employers, it certainly is in the interest of the country as a whole that workers be allowed to develop themselves to the fullest. The chance for varied and interesting work would improve life for all – even for those who mostly prefer dull routine. This has implications for the design of production and service processes and the design of tools and machinery.

As long as it is managers who make equipment purchasing decisions, there will be little incentive for suppliers to develop equipment that increases the interest of work. But workers cannot easily exert pressure on management if the technology does not exist. To break the impasse, the federal government should sponsor such development. That would be the function of this agency.

Specific projects would include the design of equipment and software that could help enable workers:

- to understand better just what their work was about, both in technical terms and in terms of the role of their activities in the larger society;
- to vary and augment their actions so as to increase their skills and knowledge;
- to divide up tasks flexibly according to preference;
- to have full control of the machinery they use;
- to communicate with other workers doing related work in distant locations;
- to control the variety, complexity, pace, and sensu-

ous aspects of their tasks to their own personal satisfaction;
● and to obtain information relevant to their work that is
now available to management only.

Two related projects are each sufficiently important to
merit special attention either as part of the assignment of
this Office, or in agencies of their own. One would be to
arrive at strategies for automating or otherwise eliminating
those tasks that no one wants to do. Concentrating automation efforts here, rather than on replacing enjoyable
work, would make most social sense. It would help make
possible a guaranteed living standard without job-holders
resenting supposed "free-loaders;" conversely, a guaranteed
income program is necessary if any plans to raise
productivity are to be fair. This point is explained further
in the description of the Agency for Technology for Full
Employment, later in this chapter.

The other special project involves a different approach to
automation. The goal here would be to equalize the status
of different jobs. The status accorded to each job is
primarily, and sometimes only, a matter of social convention, but in many cases the apparent technical necessity for
one person to issue orders to another bolsters the conventional status difference. Certain work roles, such as
secretary, technician, or unskilled laborer, are often seen as
low status because they seem inevitably to involve assisting
others in their work. Secretaries are often known as "administrative assistants"; unskilled laborers might be "machin-
ist's assistants;" and so on, throughout work situations of
all kinds. The "assistant" often really is forced to be one,
as when he or she must await and defer to decisions of an
immediate superior, saves the boss time, acts as a memory
for appointments and obligations, etc. In these cases, the
whole structure of the work situation seems to require
subservience if tasks are to be completed. But once subservience is attached to a work role, it provides a convenient
excuse for maintaining a low status assignment even for job-
holders or for parts of jobs for which it is unwarranted:

in fact, in many situations an experienced assistant acts autonomously, exercising knowledge, judgment and responsibility of at least the same quality as that of his or her putative superior.

It is difficult to overstate the consequences of unneeded work hierarchies; they promote a caste-like system, involving different ages, ethnic groups, and genders. They help maintain gross income inequalities, and they result in great differences in social power. Greater status equality could be promoted if the tasks involving deference were to be redefined, possibly automated (so that, for example, a professional could make and adequately keep track of her or his own appointments, rendering that type of assistance unnecessary.)

The Office for Gender Neutral Job Design

Most existing occupations are not gender-neutral in our culture; that is, each one is mainly associated with only one of the two sexes. When we think of secretaries, we tend to think of women, while for truckdrivers we think of men. The tools that go with the jobs are also seen as gender-specific: typewriters for women, trucks for men. These gender-specific jobs help justify a sex-segregated labor market, in which women receive considerably lower wages than men.

The introduction of new technology has historically been an occasion in which a job got assigned to one gender or the other (see Chapter 3). Therefore, as new technology is introduced, society has an opportunity to shape the job so that it is gender-neutral; that is the objective of this agency. To do this would involve careful coordination of technical factors. It also would involve attention to balancing the skills and motions involved in tasks customarily in the sphere of one gender or another.

(There is more than one way to picture true equality between the sexes. Some feminists would favor opening the role of combat soldier to women; others would view that role as representing the domination of society by inherently "male" values, and would prefer to see the importance of the role itself diminished. From the latter vantage point, gender-neutral technology may not be going far enough. Since technology has traditionally been a male-dominated activity, the field itself might need reconstituting at a deeper level, so that "female" values such as nurturance and sharing could be fully incorporated. Better defining the character of such a "female"-valued technology would be a long-term project, and might properly be the function of an additional agency.)

The Agency for Design of Occupations

Technological innovation implies the design of new occupations. The automobile helped spawn such occupations as auto repairing, gas pumping, parking lot attending, traffic policing, chauffeuring, and selling new and used cars, among others. Some of these occupations are influenced by the details of car design, some only by the general characteristic of a drivable vehicle. Most are affected as well by other technologies; for instance, parking lot attendants have a few pieces of specialized equipment (automatic gates, time clocks, ticket machines, etc.) that determine the nature of their activities in considerable detail.

The occupations related to a technology also have a reciprocal influence on the social import of the technology, and on its general usefulness. Used-car sales can involve trickery because cars are complex, and certain kinds of failures can be hidden. The aura of dishonesty that has come to cling to used-car dealers lowers the occupation's appeal to anyone taking pride in honesty, and also is a disincentive to restoring used cars to good condition.

It might be argued that used-car dealers have no worse reputations than horse traders once did, and for much the same reason. But a car is not a horse. Autos could have been designed so that faults were a little more difficult to hide, overhauls and repairs a little easier. That might have tipped the scales in favor of better dealing, and better repute for the occupation as a whole. In other cases, design influences different aspects of the related occupations: enjoyment, opportunity to increase skills, chances to meet other people, etc.

The purpose of this agency is to encourage the design of desirable and worthwhile occupations as a necessary component of technological development. To do this, it would investigate the ways that technologies influence the patterns of occupations, and would propose, where necessary, alterations in major existing and potential technologies. It would issue guidelines and offer advice to groups involved in innovation. It would seek to discover what satisfactory occupations entail, and especially what are the occupational needs of the currently unemployed. It would be responsible for raising workers' level of awareness of the relationship between occupations and technology, so that they could act to bring about what they would want.

The Office of Technology for Full Employment

Full employment, in the sense of decent, interesting, adequately rewarded work for everyone who would want it, is very far from a reality in the United States. In addition to the millions of people who are actively seeking jobs (not to mention the further millions who are stuck with undignified, boring, or poorly paying jobs), there are probably 15 to 40 million more (chiefly women long out of the workforce, teenagers, discouraged older workers, and retirees) who would seek paid work if they thought they could find something worthwhile. Unmet social needs certainly exist, so there would be work for these people to do. (The values of democracy and equality don't necessarily lead to the idea that everyone should be working; there is merit also in the idea of a guaranteed income – see below. But a guaranteed income is completely compatible with full employment in the form of socially valued work for everyone who wants it.)

In a sense, all the objections to fuller employment are social, not technical. Most people who do not have jobs nonetheless have activities they already engage in or would engage in if permitted so to do. It is a social decision that these activities are not regarded as work, worthy of reasonable remuneration. Often, a task done for oneself, one's family, or even one's community is not considered work, even though it would be if done for strangers. (This is particularly true of activities, such as child-rearing, traditionally carried out by women.) Levels of pay are also ultimately arbitrary: we could reasonably decide that fast-food workers deserve more than corporate executives because their jobs are duller and offer less freedom. Debate and struggle over these social choices should go on; yet, as elsewhere, the debate does have a technological component. Available technologies help us decide what is possible, but they themselves reflect the existing social order. Reshaping technology must go hand-in-hand with other political changes.

Almost always the immediately apparent reason for unemployment is insufficient demand for goods and services, and therefore for workers. In other words, there is no visible way for society's productive powers to be tuned to meet needs – which always do exist. This state of affairs reveals flaws in the distribution of buying power; it also reveals a lack of capacity to recognize – and to understand how to meet – needs that have gone unmet up to that point. Finally, it reveals a lack of mechanisms to organize resources so as to enable the would-be workers, with the particular combinations of talents, skills, energy, and enthusiasms they actually happen to have, to meet the demands that could then begin to emerge.

Along with conventional requirements for paying benefits, the technical arrangements of various industries make it efficient to force some workers to work overtime while others have no work. Lower-cost capital goods, more flexible machinery, better systems for deciding and communicating priorities, for consulting and planning how to meet unmet needs would reduce this apparent efficiency. Better information on plans leading to oversupplies can help reduce eventual unemployment. (Similarly, without adversely affecting the quality of benefits available to workers it would be possible to alter the form of employer contributions to remove the financial attractiveness – to employers – of forced overtime. For instance, contributions could be made proportional to total wages, without changing individual benefits.)

Monopoly power within an industry prevents prices from dropping with demand and therefore lowers the volume bought – resulting in unemployment even when the product is desirable. Therefore, technologies that reduce the "advantages to scale" are important since they can help support higher levels of competition. These advantages to scale include advantages in production, distribution, advertising, and research and development. A range of different technologies would be required, along with additional legislation to help prevent such excessive levels of competition that poverty for all concerned results – as in farming.

Unemployment within a community may be due to over-reliance on a single employer or on a single product or resource that happens to be in oversupply relative to demand. Technologies that help permit local diversification are needed. Communities could also seek greater insulation from market forces beyond their control: each could move towards being economically self-sustaining, or could work with a number of other communities to arrive at overall stability. Technologies for these purposes are discussed elsewhere in this chapter.

Current technology requires a high degree of job specialization; this promotes unemployment in two ways. First, the absence of an irreplaceable specialist often causes would-be co-workers to be unemployed. Many industries that employ numerous workers without unusual skills nevertheless depend critically on a few skilled technicians or artisans, or on operators, repairers, and maintenance workers for complex equipment. Without the specialists, work would eventually grind to a halt. Remedies might include ways of dividing jobs up so that wider ranges of skills could be shared; also, more diversified production or services might enable work to continue regardless of the absence of some specialists.

Second, employers tend to adjust the overall minimum level of unemployment to assure that each kind of specialist is reasonably easy to replace; a rough balance in each field between job openings and available – i.e., unemployed – workers is what most economists mean by "full employment." The more different kinds of occupations, the larger the total number of workers who must be kept unemployed so that an unavoidable job opening can be filled. Of course, specialists often choose not to sit idle while waiting for a job in their specialty. Instead, they "bump" less skilled workers, temporarily taking less specialized jobs. Unemployment "trickles down" to the most unskilled, inexperienced, and discriminated-against workers – currently black, inner-city youth.

Increasing the fineness of the division of labor helps each worker have a sharper sense of identity and a greater degree of job security, but it also increases the minimum level of unemployment. To solve this problem requires a sensitive balance in job design so as to maintain a high degree of individuation, ensure security and good wages, and still not make so many specialties so irreplaceable that minimum unemployment levels keep mounting.

Technical "bottlenecks" resulting in slowdowns of economic activity also give rise to unemployment. Among the causes of these are breakdowns, too few pieces of specialized equipment, loss of knowledge about how to repair or replace certain equipment, difficulty in communicating, and overreliance on a certain method or resource which has reached its limits or become ecologically unsound. Recognizing and obtaining solutions for these problems is a necessary activity of this agency.

Finally, increasing productivity without other economic changes is sure to lead to unemployment and requires a whole series of compensatory steps. If total production of a particular kind of goods simply keeps rising, it will eventually exceed not only demand but any reasonable level of need. It will also come to be out of balance with other needs. The result would be a relative increase in prices of goods, such as housing, in which productivity rises at less than average rates. Notions of supply and demand suggest that there should be a shift of employment into the less productive sector, but such shifts would certainly take time, assuming they did not meet insurmountable resistance. Meanwhile some workers would be out of work.

A partial answer would be to correlate increases of productivity so that they are as uniform as possible over different industries, and among different jobs in particular industries. That answer can only be partial because needs for a fixed range of goods are not unlimited. Therefore, a further answer would involve diversifying production to meet new needs. Even that eventually would be a limited answer: it would take time as well for consumers to adjust their tastes.

A further response to rising productivity that has long been favored in the labor movement is to shorten the average work obligation. When the work week was 50 or 60 hours, shortening it was both desirable and feasible. However, when the average work week drops low enough, a new difficulty arises. Eventually it becomes possible for some workers to have two or more jobs. Once that happens, to maintain full employment, the total number of jobs per capita would have to be increased. As long as income is based on jobs, a worker with only one supposedly full-time job would end up with a declining fraction of the average earnings, and a declining standard of living relative to the average – which is all that ultimately counts. In effect, that worker would once again be partially unemployed.

In sum, increases in productivity can lead to increased leisure time only if combined with a means of spreading income that is not tied to employment. Some technological aspects of doing this will be taken up in later sections of this chapter. Without such measures, continual increases in productivity can only mean that as a society we will keep forcing ourselves to choose between poverty and working too hard.

The Technologies for Worker Self-Management Agency

For some time, the US has been heading towards a two-caste system of employment: managers above ordinary workers. There is no clear reason for this particular division of labor, and in fact new technologies now make it easy to envision combining these functions as fully as they were in the era of family farms and small production shops. The benefits of doing so would be considerable: more interesting work for all, a wider sharing of power and responsibility, a more open, more equal, and therefore freer society. (Freer, because caste distinctions limit personal interactions and personal choices of work; they promote social tensions that further restrict ability to move, dress, or even talk freely.)

The purpose of this agency would be to oversee the developing technologies that would encourage the highest possible degree of self-management. (The agency staff itself might have to be self-managed to ensure dedication to these goals.)

Management involves such tasks as planning, coordinating, evaluating work, communicating with other institutions and customers, making agreements, and so forth. It requires access to a variety of information, an ability to conceptualize and solve a variety of problems, and related mental and social skills. Almost everyone possesses these skills to some extent, even if she or he is unfamiliar with the formal intellectual content of management. Worker self-management could be enormously facilitated by appropriate computer and communications links, and by suitable software to aid in performing the various tasks.

Among the software that would need development would be a means of recording agreements in understandable ways (a current attempt along these lines called "coordinating systems" could play this role), means of obtaining overviews of the progress of work, the status of the enterprise, the status of the industry as a whole, etc., all in ways congenial to many different cognitive styles. Another need would be a means of finding and talking with people with different specialized knowledge and particular problem-solving skills. Workers would require methods for communicating in detail with other workplaces – those that might supply equipment, for example – in many cases without actually visiting them. Ways of dividing up managerial tasks so that no one person was overburdened would be essential.

Equipment in a self-managed workplace would have to be flexible enough to permit workers to turn to management tasks, to replace each other, or to vary what is produced.

(For this agency to be worthwhile, self-managed firms would need sources of capital and financing; they would also need a friendly tax and regulatory climate. That issue could be dealt with at the time this agency is established, perhaps as the function of a companion agency.)

Technologies for improving living conditions of the poor and disadvantaged

Since technology acts as a powerful force in redistributing relative wealth, and since relative differences in wealth ultimately promote absolute differences, development must be purposely directed towards redistributing wealth downwards if poverty in the world is ever to be eliminated. The agencies described below would be especially focussed on the needs of those with low incomes and of other groups – such as the aged – requiring special treatment.

The Office for Improved Living Standards for Low-Income Households

One way to promote equality in America, given the wide disparity in incomes, would be to fix a maximum permissible income level. That would directly threaten the most powerful people; they could certainly be expected to resist reductions in their status and living standards. Another approach would be to raise the living standards of people on the bottom. The higher the level of comfort and health we can maintain at the low end, the more secure we all are. The better the minimum standard of life, the less reason there would be to undercut other workers by competing for jobs at very low pay, or to resort to crime, and the less the risk in defining one's own social role.

Arguments against improving the lowest living standards usually boil down to the idea that this would remove incentives to work, lessen the rewards for hard work, or debilitate the poor by eliminating the need to have a job. But millions of people who would work if they could do not have jobs awaiting them, and at least one important reason for that is that decent wages are not available for work they are capable of doing. Lack of a decent living standard at low incomes is actually a disincentive to work. With better standards of living already assured, the poor could afford to work even if they were not very efficient at it; simply surviving would no longer be a debilitating chore. Furthermore, not every important role in our society is now adequately attached to wages; working, in the sense of doing worthwhile things like helping maintain community or family cohesion, is not necessarily the same as having a job.

This agency would follow several strategies. It would seek means to deal technologically with poor people's own complaints. It would examine the comforts and conditions of the better off and seek to find ways around resource limitations in order to provide as close facsimiles as possible for the poor. That could include ease of transportation, entertainment devices, comfortable furniture, more varied foods. It would seek means for more effective community sharing of scarce resources. It would identify comforts unavailable to many of the poor, including lacks of: heating, cooling, safety, privacy, a variety of goods, fresh air, aesthetically pleasing surroundings, outlets for creativity, collective memory or records, etc.; it would then seek out technological means to help overcome the limitations.

The agency would seek ways to improve capacities to organize with others, to make and record agreements, to obtain essential information, and to be involved in the political process. It would identify so-called luxuries, such as musical instruments, that often contribute significantly to the enjoyment of life, and seek ways to increase their production sufficiently to be obtainable by the poor. It would coordinate its activities with such agencies as those devoted to the mass production of new technologies, universal information, and communications access (see Chapter 9) to ensure that overall strategies move as fast as possible to help the poor.

The Community Development Technology Agency

Regions of high poverty, both urban and rural, are to be found throughout the United States. It is commonly agreed that these levels of poverty are not due to either lack of desire to find work or lack of ability. Instead, there are no willing employers present. Continual debate on the correct method to close this gap has failed to lead to a widely accepted policy, or to a serious, sufficient government commitment to ameliorate these conditions. The favorite proposal of the Reagan administration has been so-called enterprise zones, in which businesses would receive incentives such as lowered minimum wage requirements in return for locating in these areas. In effect, the enterprise zones would function like duty-free export zones of Third World countries. This approach not only undervalues potential workers in poor communities, but it pits them against workers in other areas – who might end up losing their jobs or having to accept lower wages. This "solution" therefore will help increase ethnic and geographic tensions and divide potential allies.

As the plan is envisioned, most of the businesses to be attracted to the enterprise zones would be fairly large firms that would use the new facilities to produce goods or services for higher-income people living elsewhere. Goods bought with the new incomes would mostly come from outside the zones, and profits from the businesses within the zones would also quite often go elsewhere. Therefore, the new jobs would at best lead to a minimal improvement of living standards within.

What is missing in the enterprise zone proposal is any awareness that the poor have enormous needs as well as the capacity to work. A more reasonable proposal would focus on these needs, so that enterprises established in poor districts would produce goods and services for the poor. With economic circulation within the region, there could be a substantial multiplier effect: each job could lead to several more and the level of goods and services could continue to rise. That outcome would be enhanced if profits were to be reinvested in the area. Since overall economic activity would rise, outside workers wouldn't have to fear for their jobs. This model of entrepreneurial development for home use is closely tied to models of successful development in other countries. Success would not so much require attracting outside capital as helping indigenous entrepreneurs to get started. Additional help is needed in developing an internal "banking" or bartering system; this would amount to a replacement, for purposes of exchanges with one another, of the money that poor people so obviously don't have.

Technologies to support such community-based entrepreneurship would include the special banking or barter system; small-scale, flexible production and service technologies; tools for rehabilitating buildings as workplaces; help with organizing tasks; help in converting ideas into feasible proposals; help in adjusting work to initial skills; production processes geared to allowing on-the-job skill development; means of discovering needs and capabilities; recycling and remanufacturing processes, etc.

The Agency for Production at Appropriate Scales

To equalize power in the world, it is enormously important for each community – however defined – to retain a considerable degree of economic autonomy (see Chapter 3).

Most people would identify themselves as members of several different communities; among these, the smaller communities are generally embedded in some way in larger ones. The larger a community, the greater its degree of potential autonomy, so starting with the smallest community of which any person is a member, and going outwards to ever larger communities, the degree of autonomy could increase with each step. Within that framework, how much autonomy is possible at each level depends on what technologies are available.

This agency would have the function of devising technological strategies for increasing possible autonomy at the smallest levels. In order to do this, the agency would attempt to establish alternative models of goods production. One goal would be to maximize possibilities for sharing of goods, recycling, and rehabilitating ("remanufacturing") of existing goods.

Since a small community cannot produce everything, and since autonomy implies choices, several ways of dividing up what can be produced at each scale are needed. The agency would attempt to design alternative sets of goods – permitting a full community life – consistent with different intensities of production on various scales. To accomplish this, the agency should also plan for sufficient variety of occupations and skills at the smallest levels, so that there could be a meaningful, definite role available for each member of a community.

Some goods, such as ocean freighters, clearly cannot be

produced in small communities. But, in a world of community-based production, many such goods may not be needed to the extent they are currently. An important part of the strategy would also be encouraging substitutions for goods which would otherwise have to be imported. For example, locally producible crops, alternative materials, energy conservation, small-scale machinery manufacture, and biotechnological methods of drug and chemical production can all reduce import requirements.

The agency would devise communications and related technologies to permit and encourage such activities as barter for kind, to help increase the potential for economic autonomy within communities, as well as for aiding community self-definition. Further, it would help establish criteria for communication among communities, including exchanging key technological information, that would help autonomous functioning at the material level.

This example illustrates the difference between technologies that emerge from institutions following the values of the military or of corporate profits, and the very different technologies that should emerge from this program. Materials substitution technology would not concentrate on replacing the materials that are scarce internationally, are "strategic" for some military purpose, or have been patented and are therefore not producible without hard to obtain licenses. Instead, it would seek substitutes for any material that might impose a barrier to any community's efforts to meet its own needs, with several possibilities for achieving any specific end. A fully developed alternative materials technology would allow each different region and different climate its own mode of operation.

The Agency for Technology for Special Groups

Special groups in society – including the elderly, the physically handicapped, anyone who differs from a rather rigid norm in size, strength, mental abilities, etc. – would be able to cope better, were technological efforts focussed on their particular needs. This agency would do that, with the goal of allowing members of these groups to live fully, while requiring the minimum in burdensome attention from others. The agency would not only examine possibilities for helpful devices, means of communication, etc., but would also consider how the community as a whole could alter its practices so that the members of special groups would face fewer barriers to full participation. The agency would seek to develop technologies that are as affordable and as minimally disruptive as possible.

The agency would examine all aspects of life for each special group. It would also seek to identify other groups who could benefit from this kind of attention. Like other agencies proposed in this chapter, it would carry out all its activities under the direction, as much as possible, of representatives of each special group.

Agencies promoting democracy, cultural equality, and community enrichment

Technological change has a marked effect on power relations, on cultural possibilities, and on the possibility of community life. A democratic society must continually provide for augmenting capacities for democratic involvement if the direction of technological development is not to lead to their atrophy. A central part of democracy is the ability to engage in effective public dialogue. That ability in turn is predicated on cultural possibilities. Assuring cultural equality must therefore also be a goal of a sound government technology policy. Democratic possibilities also will be limited if community life becomes difficult; a single, nationwide arena for dialogue is necessarily one in which only a tiny percentage of voices could be heard. Hence, assuring some kind of rich and worthwhile community life is also a vital requirement for technological development. Ultimate responsibility for all these areas must rest with the government. The following agencies would each deal with some facet of these broad obligations.

The Agency for Technology for Improving Childhood

A few decades ago, the US was seen as a child-centered society. This is certainly not the case any more. With so many fathers neglecting to pay child support, with the large number of poor, single mothers, with middle-class women delaying having children until they can interrupt their careers, with the decline in children's movies and even in television programs specifically for children, with declining support for schools, shrinking federal grants for child nutrition and related programs, growth in (reported) child molesting and kidnapping, housing complexes that do not permit children, children and their parents face a rawer, more hostile world.

More single parent families and, simultaneously, more women working means parents being forced to leave their children alone at least part of the day. This is especially nerve-wracking because of increased reports of kidnapping, child molesting, and even child murder. Adequate childcare is one obvious partial answer, but, especially for school-age children, it seems an unlikely prospect. In any event, there will always be times when there is no childcare available.

There is a deeper dilemma as well. Most children's images of the world are now formed by images intended for adult (or at least adolescent) consumption on television. This is a sharp break with the trend, in prior generations of the modern era, that lengthened the protected period during which children could master the skills and responsibilities needed to function as equal adults in the confusion of an industrial, market-based society. Yet, while they are losing this protection in terms of images, children are being increasingly isolated from the world of working adults.

The technical environment of the workplace is one reason children are excluded: toxic fumes, radiation, dangerous machinery, as well as delicate electronic equipment, and "clean rooms" for microscopically precise work, all help place high tech and ordinary factories, laboratories, and offices out of bounds. Urban designs which segregate residential areas from the rest of life create another barrier, which is only heightened by the limited availability, frequent breakdowns, and loss of safety of mass transit systems. The growing percentage of children who live in poverty are particularly affected. All these trends taken together present dangers of an early apparent sophistication masking a lack of capacity to comprehend or deal with difficult realities.

This agency will seek ways in which technologies may be modified to help make it possible for children to avoid isolation, by maintaining contact with their parents, by safely moving through the adult world, by increasing their access to means of communication, etc. It would seek to develop portable, child-usable communications devices suitable for such purposes as signalling to a specified adult for emergency aid, transmitting the child's location in an emergency, and helping children themselves know where they are. Related communications technology might facilitate cooperative childcare arrangements by permitting the parents in charge to call for assistance in mild emergencies.

The agency would also help reshape the landscape in residential areas, even poor ones, by designing alternative shared workspaces that could be safe for children to enter, and by devising patterns of interesting and worthwhile adult work in which occasional distraction by children would not cause notable stress. It would develop technologies to ease the isolation of single parents as well, to reduce the stress of childrearing, and to help people living in the existing housing stock (especially in suburbs and poorly designed housing projects) overcome what amounts to physical isolation. These latter technologies might include public transportation systems designed for low density settings.

The Agency for Heightened Democratic Involvement

If democracy is to remain meaningful as the world continues to become more complex and interconnected, and as the number of decisions affecting the average person continue to grow, then the possibilities for democratic involvement have to continue to be extended as well. Democracy entails more than the right to vote. There must be public spaces where issues can be presented, discussed, and negotiated, where political groupings can form, disperse, and reform. Individuals need access to information, opportunities to seek out wisdom, and forums to express their opinions.

Most important, the existence of democracy is a statement about the human world, not only about the apparatus of decision-making. For democracy to be more than a ritual of voting, the world has to be stuctured so that small-scale decisions matter. A world of giant power plants, strategic missiles, and industrial planning at the global level is not a world in which democracy can flourish.

All these aspects of democracy have implications for desirable technology. While traditional public spaces such as open city squares, cafes, and meeting halls remain important, in our highly dispersed society we need additional kinds of public "spaces" that are not real places but are means for people who share a particular concern to communicate and coordinate over distances. There should be facilities for like-minded people to locate one another, and avenues for new groupings to make their presence known. We need new means for people to offer partial and conditional proxies to representatives chosen for particular issues; and we need technologies that permit fairly small groups flexibly to coordinate responses to any sort of social decisions affecting them. These facilities would be open to both new, informal groupings and existing organizations, including political parties and labor unions.

To create these technologies is a task for society as a whole. The magnitude of the challenge is probably greater than that for any other agency proposed here. The primary responsibility of this agency would be to determine what sort of technologies necessary to achieve these goals are least well developed, and to concentrate its resources at those points. It would also monitor the overall state of useful technologies, proposing ways of combining them to help enlarge the democratic system. Further, it would have the duty to warn of currently emerging technological perils to democracy.

The Agency for Technologies for Alternative Markets

The central fiction of modern economics is the existence of the marketplace. In reality, marketing is a complex process involving: varied modes of organization; access to channels of communication between organizations; access to mass media; systems of producing, packing, transporting, storing, and locating goods; and symbol systems associating the names of particular goods with the more free-floating desires that people actually have.

In the market squares of premodern economies, to participate as a seller one had to have a booth. Space was limited, and therefore access to the market was also circumscribed. The same is true today. The effective channels of communication, and people's abilities to remember and associate symbols with desires, also are limited. The powerful corporations that control many of these channels are not particularly interested in allowing new ones to be born. The greatest difficulty most new producers of goods, services, or ideas have is gaining entrance to the market.

Radically new kinds of products open up the market for new companies, because they entail new conceptual categories. When this happens, the companies that manage to gain a toehold typically attempt to exploit it by spending all they can to exclude others from competing. A good recent example is the field of personal computer software. The costs of designing new software or producing the discs needed to distribute it can be dwarfed by advertising budgets. The few highly profitable early successes in this market try to monopolize shelf space at dealers, to dominate the thinking of dealers and customers, to raise the cost of developing competing software by accustoming potential buyers to expect explanatory books, video tapes, and other amenities that may or may not actually be very helpful.

Computer software also illustrates the possibility that very different types of access to many new goods are constructible. Such software may be fed through the telecommunications system to numerous outlets or even to individual users at low cost. More experienced users of the software can advise or instruct newer users. Improvements or modifications can be fed into this alternative market by whoever makes them.

In the case of software for personal computers, the alternative market actually exists, albeit in rudimentary form. "Users groups" exchange "nonproprietary" or "public domain" software (software which no one claims to own) intended for a wide variety of purposes, including many of the same ones served by better known "proprietary" software. There are difficulties, in that under the current system, no one is required to take responsibility for reporting errors or problems, much less correcting them, for the nonproprietary software. There is no direct compensation to the producers of nonproprietary software beyond the good feelings of benefiting society, while successful producers of proprietary software can gain substantial wealth and power. New users often only find out about nonproprietary software by luck. Meanwhile they are deluged with information about proprietary wares

There are a series of social as well as technical issues involved in creating alternative market forms and new methods of access. Some of the social issues are discussed in Chapter 10. The function of this agency would be to deal with the more technical issues that also need attention; these would include creating effective alternative communications and distribution channels to allow wider market access. The agency would seek to devise communicative forms that would resist efforts at monopolistic dominance, that would not have to be supported by advertising revenues, that would permit users and consumers of the products and services involved to exchange information, to help one another, and to reach the producers with requests and advice.

The technology devised should permit some reasonable degree of compensation to producers. This will involve new modes of record-keeping – designed not to endanger the privacy of the users. The technology developed should be extended to as many types of products and services as possible, should allow direct communication among designers, producers, and users. It would allow channels for presenting reviews or critiques of products – channels so designed that they could not easily be misused by firms intending covert slander of rivals' products or unwarranted praise of their own.

Technological developments could also help groups such as farmers, who are now subject to vagaries of international markets and commodities exchanges beyond their control. A system of communicating ranges of probable yields and anticipated demands could possibly be tied with methods of negotiating increases or reductions in plantings more equitably than at present, while lowering consumer prices.

At the other end of the geographic scale, small businesses now find themselves in competition with franchise chains that often provide less good and less community-oriented services but, due to the nature of mass media advertising and large-scale distribution systems, are often able to dominate markets. It is possible that new, easily usable forms of local media and association could help offset such advantages to franchises.

The Agency for Technologies for Widening Cultural Equality

Our society puts a premium on certain forms of knowledge that amount to a definite culture. Growing up in a middle-class, literate family, having access to certain facilities – currently including computers, but also books and libraries – enhances prospects in a highly competitive job market as well as in the political arena. Inequality also results from a testing system, active throughout schooling, that is organized on the explicit assumption that people should not only be tested but ranked, so that only some can be accepted as performing adequately. Those who do not perform well on such tests, possibly through having grown up in a culture substantially different from the dominant one, learn that they are unequal, that they should lower their expectations, and that they will never quite understand the world in which they find themselves.

A more egalitarian system would acknowledge differences in performance or culture, but would not treat such differences in terms of success versus failure. This agency would seek to lessen the effects of cultural differences. It would attempt to restructure communicative devices, systems of knowledge, and other sorts of tools to be adaptable to a wider variety of ways of perceiving the world. For example, it would build on the potential of computers to allow thoughts to be presented in terms of a wide variety of still or moving visual images, or in terms of non-standard vocabularies, grammars, or dialects, languages other than English, or alternative world pictures. Through this, it would enlarge access to knowledge, e.g., mathematics, heretofore considered the special province of a small elite. While such an endeavor cannot make everything intelligible to everyone, it can continually enrich understanding and ability for almost all. Limitations to comprehension that turn out to exist according to one alternative approach might disappear with yet another.

The Agency for Assuring Wide Literacy in New Media

New as well as old media are fully comprehensible only to those people who have an opportunity to "write" as well as to "read" in them. For the newer media, this requires a deliberate agenda for developing technologies relevant for "writing" and making them widely available. For instance, inexpensive television cameras, for community use, could have been a priority for development years before they were. Currently, computer graphics capabilities even remotely approaching the complexity of those used to create images in science fiction movies are inaccessible to the public at large. Likewise, most children have little chance to learn what computer programming is really about. This agency would be responsible for developing tools to promote greater literacy in all new information media.

The Agency for Intelligible Technologies

Technologies such as microelectronics are difficult to understand, since they operate at an invisible level. Related devices often come equipped with the warning, "Do Not Open. Danger of Shock. No User Serviceable Parts Inside." The purpose of this agency is to promote designs of technologies that are intelligible to their users or simply to the curious. It would start with the most commonly used, but nonetheless unintelligible, technologies, such as electronics, and spread out from there. Where small scale (e.g., microchips) makes direct observation of the working of gadgets difficult, the agency would devise alternative means to make the manner of their operation understandable.

As much as possible, elements of this system of explanation would be included directly in products. For common technologies, the agency would prepare simple, easily reproducible packages of explanations. It would collaborate with the two agencies just described in doing this in a variety of cultural symbols, accessible to a very large fraction of the total population.

Additional agencies

The Agency for Technological Aid to Village Development

Current patterns of Third World development are creating a class of unemployed, dispossessed rural poor who are forced to migrate to urban areas where most remain in terrible poverty, and are fed only through further impoverishment of the villages. This agency's aim is to aid in improving village living standards so that this cycle of hopelessness can be overcome. Working in tandem with the Agency for Production at Appropriate Scales, it would offer technical advice to help villagers preserve their basic culture and improve individual farming practices by minimizing their dependence on external sources – e.g., for fertilizer. It would then help them develop village and regional production so as to improve living standards and limit burdensome toil.

This agency would have to proceed with great care, since traditional village practices often have multiple functions, so that a seeming improvement in one particular function may easily be harmful for others. For instance, improving cookstoves (a favorite topic of appropriate technologists) is not simply a question of lowering fuel consumption: light, space heating, smoke (for repelling insects or flavoring foods), and ash production – along with symbolic values – all matter.

Further aids to even development would include lowering costs for transportation and communication between city and country, and between villages, so that goods produced in cities, and production itself, could reach the countryside as early as possible.

The Agency for Technological Balance

Even when technological efforts are attuned to promoting democratic and humane values, the overall results can be unbalanced: certain values might be overemphasized by a whole range of technologies, while related values are insufficiently supported. Thus, technologies that keep hospital patients alive in the narrowest sense are overdeveloped in comparison to technologies that might help patients live out the ends of their lives in dignity. There are several ways of dealing with such imbalances, ranging from reducing expenditures and altering practices in the overdeveloped areas to identifying and increasing attention for hitherto neglected values. This agency, working in close contact with the various commissions for oversight of the impacts of technology, would follow the latter course. It would identify neglected values and propose corrective agendas to be carried out by it or another agency.

9 A model "Public Technologies Program"

Providing an "infrastructure" for the life of the nation is one of the recognized responsibilities of government. Sidewalks, roads, canals, harbors, airports, libraries, agricultural extension services, schools, sewer systems, and irrigation systems are some of the things regularly maintained by governments from municipal to federal levels. While some of these public works are operated on a fee for use basis, others, such as sidewalks, are commonly available at no charge to the users. As new technology leads to cultural changes, such infrastructure must be expanded to new categories of facilities and services if any sort of social equality is to be maintained. Otherwise, the benefits of new technologies will inevitably accrue preferentially to rich corporations and individuals.

Public works are not only egalitarian, they are generally far more socially efficient than relying on private intiative would be. By making the best use of resources, they enrich the life of the nation or community as a whole; conversely, without them, we would all be impoverished. Private roads or sidewalks would likely not provide links for poor or small communities; even in wealthy communities they could only be adequately maintained through private organizations operating as quasi-governments. In many instances also, still further social efficiency is gained by having such facilities available for use free of charge, rather than taking up the public's time and resources in the practice of paying and collecting tolls and fees.

This model program, then, is a proposal for updating essential, worthwhile, and time-honored government activities by inaugurating new categories of public works that new directions in technology make possible. It would lessen scarcities and lower inequalities in opportunities for living well by increasing the variety of satisfactions that could be made available to all.

Public works are often costly, and these would not be exceptions. Each project would make some sense in a pared-down version, but obviously could take on an altogether different character if more lavishly supported. Since it would be inappropriate here either to offer a definitive budget or to list the possible variants, each proposal is presented as more or less "open-ended," subject to virtually infinite elaboration or variation.

The Universal Access Information Network Authority

Improved communications and widened access to information of all sorts would enhance equality, enrich our culture and our social lives, benefit our economy, and permit a far fuller democracy. At present, the rapid influx of new information and communications technologies creates a worsening division between those who can afford the latest means of expression, organization, and obtaining information, and who therefore learn sooner how to use them, and those who are several steps behind. Because information is not used up as it is used, this is a resource that, more than any other, could help increase equality without imposing restraints on anyone. In many cases also, communications media can replace transportation, saving resources, land, and energy, and lowering pollution. Overcoming unnecessary limits on access to the new capabilities deserves high priority.

New information technologies include new forms of transmitting, receiving, carrying, displaying, recording, obtaining, and utilizing information, whether written, oral, audio, graphic, numerical, visual, still, or moving, or of any other kind that is transformable to one of these. Examples include satellite communications, fiber optical signalling, digital and high resolution television, laser discs, supercomputers, local area networks, microcomputers, cellular radio, computer data banks, electronic mail, etc. The largest category of growth, however, is in what has now come to be known broadly as "software," including computer programs, information stored in data banks, the contents of films, etc. The size and complexity of these two lists show why only substantial corporations, government departments, and the very rich are likely to enjoy full access to information and communications. What is only a little less apparent is that all these different systems are being made increasingly compatible with one another, so that they could form one huge network. If properly designed and implemented, this network could provide all the services to almost everyone.

This act would establish a nationwide Universal Access Information Network Authority, with the mandate to ascertain what technical developments would be needed to make a high performance system likely to meet all the nation's two-way information needs over the next century or so. This system would have to provide:

1 Convenient access for every person who wanted it. This would include access from home, work, or public places.

2 A signal-carrying capacity at least sufficient to permit each person's sending as well as receiving the equivalent of "high resolution" television. This capacity would make it possible to conduct two-way television conversations, transmit pictures of objects or high quantities of data, and make possible a wide variety of consultations between households. (The most obvious value of two-way television is that most people are capable of far more effective dialogue if they can see and be seen by their interlocutor than if they have to rely on sound alone; making this capability widely available would by itself vastly increase possibilities for democratic involvement, new forms of

community, etc. It would be especially valuable for everyone who is not in a position to travel great distances frequently, which is the situation of most of the relatively disempowered groups especially.)

At present, of course, the major means of signal transmission from households is by means of ordinary telephone cable. Without government intervention, this system is likely to be replaced piecemeal, with poorer households receiving completely inadequate systems. New installations, if required, should provide enough added capacity to obviate the need for augmenting them for many decades; the requirements for high resolution two-way television would seem ample for most other conceivable purposes as well. Installing this signal-carrying capacity might prove to be the most expensive part of the whole undertaking; in any case, it is an area in which only government involvement can assure equality of access. Appropriate sending and receiving equipment would also be necessary, of course, but to be useful would require a lower level of coordination.

3 Access to as wide as possible a variety of information that can now be stored in digital form, or can be expected to be so storable in the next few decades. This would involve access to materials in libraries, including audio, video, and computer software libraries, as well as government files of a non-personal nature. Funds would have to be available for maintaining and improving such collections, including the classification systems by which items could be located. Since information now in libraries is subject to rapid physical decay, the program should include technologies to assure permanence; means to preserve public documentary information (for instance, memos or drafts), which once was routinely saved in archives but now tends to be produced on computers and then rapidly erased, should also be developed, as preserving such historical records is a necessary component of maintaining democracy.

4 Capabilities for "networking," forming groups, political organizing, electronic mail. This poses requirements on the switching system to be installed in the network,

and on amounts of publicly available computer-memory capacity.

5 Capabilities ensuring the right to privacy for all individual users. (This would include such rights as the right not to be added to lists, the right not to be spied upon by government or large organizations, the right of access to and review of any files that include information about oneself, provided that that information was not kept for the purely personal use of the filer.)

6 The capacity for anyone to be able to add information to publicly accessible data banks.

7 The analytic capacities of fairly large computers available for public use. Additional computing capacity could be added to satisfy demand.

8 Compatibility with reasonable possible improvements at a reasonable price.

The Authority would attempt to determine the specific, feasible technical developments that would permit installing a network of this type within about a decade, at a reasonable cost. What cost would be reasonable is matter for debate; the tremendous improvement in public life and opportunities this system would provide might justify expenditures amounting to a sizable fraction (say, a few per cent) of the gross national product for each year it is under construction.

The Authority would publish the specifications for the system and for the parts requiring further development or research. Development could proceed by several routes. Corporations would be encouraged to compete in developing the various parts of the system at the prices suggested by the Authority or less. They would be assured that success would mean substantial sales. Those parts which it seemed no corporation would be likely to undertake on its own might be subject to bidding or could be developed directly by the Authority.

The Authority would also decide on a flexible overall plan for putting the system in place as soon as possible. The most expensive part of the system would almost certainly be connecting individual households. The first priority in construction might be building these networks in the poorest neighborhoods and areas in the country. Since doing this would increase the value of housing in these neighborhoods, the success of the effort to improve the status of the poor would depend on carefully preserving their rights to continue to live in these neighborhoods.

The government would subsidize or completely under-write the costs of installing the system in poor neighbor-hoods. In others, it could pay a pro rata share.

The Authority or a separate commission would attempt to assure that information of all sorts would be available, in understandable and usable forms, to all users of the Universal Access Communications system, as soon as poss-ible. As necessary, it could establish orders of priority in readying data banks, could establish protocols whereby existing data banks may be added to the system and new ones formed, and could call for the collection of new categ-ories of information if a clear use were evident.

The Agency for Promoting Mass Distribution of New Technology

As emphasized in Chapter 3, rapid technological change leads to rising differences between the well-off and the poor. The more a technology helps augment personal (or class) power, the more its being expensive provides a barrier to equality. Since recent innovations tend to be expensive, power imbalances grow. Social justice and preserving democracy require efforts to close this gap. This means shrinking to the minimum the time interval between when technologies that increase power are first introduced and when they are readily available.

The agency proposed here would begin by identifying those technologies that, under normal operations of the market, can for some time be expected to benefit only the well-off and that either would be likely to be a prop to social or political power or could substantially improve the quality of life.

There are a wide variety of possible reasons that these technologies may not be available to the relatively poor. The most obvious reason is high prices, which in turn also have more than one cause. And since initial production facilities obviously must start with limited capacity, innovating corporations often, and not unnaturally, try to maximize the return on their development investments by charging as high a price as they can get. Since initial production facilities are often quite experimental, they rarely are highly efficient, or even particularly cost-conscious. The more complex and unusual the product, the more the supplies that go into it also will be produced under similar conditions.

The new products are also likely to be marketed very selectively. Since the better-off can afford to take expensive risks, they are the logical initial targets of advertising. If supplies are limited, so will be the shops and mail-order houses that sell them, and these too are likely to focus their attentions on only the best potential customers. With the fairly well-off as the obvious target for sales, the product will be designed and refined to fit the needs and aptitudes of this group. Every aspect of the product will make it seem unsuitable to the needs or cultural style of the relatively poor, and the low interest in the product such communities show as a result will be taken as further proof that the appropriate market remains the better-off.

To change this process, the agency would probably need to break every link in the chain of restrictions. The very existence of the agency would help promote new kinds of private sector innovation, geared to wider needs. The agency might have to choose among these, and if so, it should probably assure diversity by having a number of independent selection procedures. For the technological innovations it selects, it may have to offer some sort of guarantee of large-scale sales in return for rapid expansion of production capacities. It might have to organize or underwrite efforts to devise more efficient production methods, perhaps through public, regional, pilot production facilities; this part of the program should be run in close collaboration with workers and with the Office for Improved Work Quality.

New distribution channels would also be required, and there may be a need to support technological modifications that would make the resulting product of more direct use to the poor. The agency may also have to estimate plant capacity needs, help set up regional industries, and promote necessary research.

Public production facilities

Another, obviously controversial step the agency could take would be to establish public production facilities. Beyond the basic rationale behind the agency, there are two additional functions these could serve:

1 In Chapter 7 it was suggested that the diversity promoted by competition between firms comes at too high a price: the threat and reality of unemployment, declining incomes, and loss of community control. If government provides a set of facilities and services in which some development and even production could take place, a real, but sheltered competition without constant threats to incomes and community status might be possible.

2 The following chapter discusses the ways in which the intellectual property laws currently operate supposedly to encourage creativity, and suggests the need for alternative mechanisms. Publicly sponsored production facilities, especially if suitable for pilot production, are again an obvious possibility.

The size and scale of the facilities could vary widely, depending on the degree of public support. There should probably be an emphasis on multifunction factories designed to supply goods for communities. These could be supplied with a diverse set of tools and equipment that could be adapted fairly easily to a whole range of differing products.

The facilities might be made available on a variety of terms – for instance, self-managed worker cooperatives might be permitted free use of them; small, for-profit companies might be expected to refund a portion of their profits to the agency; in other instances, non-profit production organizations, possibly subsidized further to obtain parts and materials and to pay workers, might be necessary and appropriate.

The factories and workshops themselves might be new or largely recycled. They could be preferentially located within low-employment communities, grouped together or dispersed, according to circumstances.

The National Infrastructure Renewal Commission

As a society, we have generally been better at building anew than at repairing or maintaining what we already have. Years of neglect have been particularly wasteful and damaging to much of the public infrastructure. The existing contracting system makes it much easier and more lucrative for private companies to undertake large-scale construction than to be involved in piecemeal repairs, and they usually lobby accordingly. Funds for timely repairs or routine maintenance do not get voted with any consistency, so that the public works departments responsible for maintenance often have never learned how to perform these tasks even when there are funds. Still less, proportionately, are there adequate research and development programs directed at the issue of repair and maintenance; for example, hardly any funds are available to obtain new insights into the properties of materials that might be important for saving existing road surfaces, or for designing repair procedures so as to minimize disruption of community life.

This commission would have to decide what bits of existing infrastructure deserve the most attention. This decision would have to involve not only environmental and resource considerations but careful judgments on alternative

180 / Reinventing technology

future plans for economic and social life in the regions in question. To be worthwhile, those judgments would in turn require democratic participation. The choices made would then determine funding for research and development, as well as for actual repair and maintenance projects.

The National Skills Conservation Commission

One of the negative consequences of technological change is that as old techniques are superseded, the skills associated with them are often lost. That is a real loss, not only because such skills may have provided pleasurable work, but because they might have proved of renewed value to society as a whole if circumstances, such as the relative prices of raw materials, were to change. Sometimes the road of technological change leads to costly deadends, as appears to have been the case with ever-larger electric power stations, for instance. Technologies can prove unexpectedly complex, have unanticipated drawbacks, or rely on a level of resource use that comes to seem profligate. Often poorer communities are in older parts of cities, which are especially subject to fall into disrepair, in part because no one remembers how to fix or maintain older buildings or equipment such as elevators or plumbing. Finally, there is much value in a society's retaining an understanding of its past, if only to avoid repeating the same blunders; that requires retaining the experiential knowledge that skills represent.

The commission would recognize existing skills as valuable national heritages that could not easily be regained once lost. It would seek to catalogue existing skills (this catalogue might properly not be limited to technical skills but also include social skills, such as the skill of training apprentices, and perhaps others, such as pretechnological crafts, as well). The commission would instigate a system to warn of those that are disappearing. It would further devise means of skills preservation, including supporting apprenticeships in endangered skills, honoring people who keep alive old skills, and, for skills that no one can be persuaded to maintain, finding the best possible means to record them fully as actually practiced. It would conduct research to revive lost skills. And, finally, it would establish local boards to search out regional skills.

10 Defining the climate for innovation: A model "Intellectual Claims Act"

Introduction

This chapter concerns a revision of laws and policies affecting trade secrecy, patents, and copyrights* – or more broadly, what is known as "intellectual property." The intellectual property laws are the major means by which government influences and regulates the climate for private sector innovation. The system of intellectual property, among other things, determines who can make what; for this reason alone its importance would be hard to overestimate.

More than for the other policies discussed so far in Part Two, details are what matter in determining the effect of the intellectual property laws. It seems appropriate, therefore, to present an alternative very roughly in the form of model legislation– a proposed "Intellectual Claims Act." Other policies influencing private sector innovation, such as tax incentives for research, will be taken up briefly at the end of this chapter.

There are two different justifications for our current statutory system of intellectual property. The first is that creators, like everyone else, should have a right to benefit from the

---

* I include copyright since that law now covers technological entities such as computer programs, and also since all manner of literary and graphic materials are now subject to reproduction through an ever-increasing variety of technologies.

fruits of their labors. But just what are the fruits of the labor of invention? If someone invents a new kind of potato peeling machine, does she or he deserve a laurel wreath, one of the machines free for personal use, a lifetime supply of peeled potatoes, or royalties of five dollars for each and every time anyone uses one of the machines?

If what the right should amount to is collecting royalties (how much?) on every potato peeling machine sold for seventeen years, then the existing law is exactly correct. But this is of course a circular argument; that is the right because that is what the law is. Changing that law would not violate the right but would merely redefine it; the right to intellectual property would be an utterly meaningless concept without a state to enforce it. In fact, the legal theory behind current law has nothing to do with such a right.

The legal theory, which is the other justification for the intellectual property law, is purely a policy consideration: We as a society will benefit from innovation and therefore choose to grant patent and copyright protection as an incentive to innovative activity. This is the Constitutional basis for patent, copyright, and related forms of intellectual property (see Chapter 5). In current theory at least, the incentive is needed to encourage much besides initial creativity. In the case of patents, for example, innovative activity goes far beyond actual invention to encompass product development, production method development, plant investment, initial production runs, establishing a distribution network, and advertising the new products.

If intellectual property protection is a matter of social policy, and not simply a right, then that policy must be weighed according to the entirety of its effects. The more wealth takes the form of intellectual property, the more crucial this evaluation becomes. Perhaps because it seems technical, discussion of intellectual property in recent years has been left mostly to those who would benefit directly from extensions of the law to new areas. With a few exceptions, such as the Supreme Court decision to permit video-taping of television programs at home, other public interests have not been addressed. Much ground has been lost; the time has come to regain it.

Considering the scale of innovative activity at present, it is not even obvious that it is good policy to encourage further innovation. There must be some rate of introducing new technology beyond which we will have social chaos. A policy of promoting innovation ought to be geared to the capacity of our society to absorb change in an equitable and broadly enriching way. Further, there may be certain kinds of innovation that are much more important to encourage than others.

A sizable share of innovation would certainly continue without the incentive of patent or copyright protection. This is evident in areas not covered by the law. For example, it is difficult or impossible to patent chocolate chip cookie formulas, and yet this has been a growing and innovative business. The kind of chips currently taken more seriously – microelectronic integrated circuits – also underwent feverish innovation for years, despite the lack of clear legal protection for specific circuit designs until 1984.

Instead of encouraging beneficial innovation, then, the effect of the intellectual property laws may be more to aid in the concentration of wealth and raise profits. The firms that benefit can then use their increased power to further prevent competition from outside innovators and thus increase their control over the direction of innovation as a whole. State power ought not to be used in this manner – to strengthen the already strong. Rather, it should help to promote greater equality. In conferring wealth on innovators, it is well to remember also that their achievements, no matter how great, rest on a basis of prior inventions and discoveries of society as a whole – many of which have already been paid for directly by the federal government.

One central assumption of the patent system is that in return for an exclusive license, the patenter will reveal the details of the invention to the public, so that the same techniques could be used for other purposes, and so that when the patent expires competitors will be in a position to produce the same product. In reality, the patent system often fails to fulfill its informative function because patent "disclosures" are not required to obey reasonable standards of informativeness; for instance, they need not have topic sentences indicating the general character of the invention, indicate their connection to previous inventions, nor be clear to nonexperts.

Patents are often presented in vague, confusing form, both to extend the claims as widely as possible and to throw potential competitors off the scent. Sometimes this enables the innovators to file subsequent, more specific patent applications that serve to extend the total period of patent coverage, in some cases for decades.

Simultaneously, a patenter can resort to trade secrecy (see below) as a means to restrict further the ability of potential competitors, including current employees, ever to compete. This narrows the channels for inventiveness and makes it impossible for informed public debate on the merits of an invention to occur soon enough to have much effect on design and production plans. Sound policy would move toward greater openness.

A more repressive aspect of current law is in enforcement of intellectual property rights, which are rights to ideas rather than things. Unlike other kinds of property, intellectual property in principle can be "stolen" anywhere, even in the "thief's" own home or business. The likelihood of such "theft," and therefore the intrusiveness required for effective law enforcement depends on both the nature of the innovation in question and the ease of available means of copying.

For instance, at present, to attempt surreptitiously to copy a current model low-priced car would be absurd. Any economic scale of copying would require such a high volume of public sales that detection would pose no problems.

But two trends increase the attractiveness of "stealing" intellectual property while making the steps needed for effective prevention or detection much more questionable. One is growth in the number of intellectual property items for which prices are high compared with direct production costs. Examples are motion pictures, books, and many categories of high technology, including microelectronic chips and software.

The other trend is that technologies of reproduction, such as photocopying, video, and copying of computer diskettes, are becoming more versatile and accessible. On the whole, this is a tendency to be applauded. It would be essential for widening access to political power or for diversifying economies at the community level, as called for in Chapters 8 and 9. But the two trends taken together mean that, as currently defined, intellectual property rights can be protected only through increasingly serious limitations on expression, or violations of privacy, or both. Protected categories of information are now often stored in computer systems in such a way that they are accessible over telephone lines, so that one may now be considered to be committing theft by telephone. The law thus invites extremely intrusive enforcement procedures, incommensurate with well established notions of privacy; it becomes a perfect excuse for far-reaching snooping into personal beliefs, etc.

Violating privacy and freedom of expression in order to encourage novelty seems perverse, yet encouraging novelty is the purpose of intellectual property laws. If anything, this repressive enforcement would be likely to have a chilling effect on creativity of all kinds. A sound policy would not permit incentives to innovation to involve granting rights that can only be enforced so invasively.

For many inventions that might be difficult to protect under current intellectual property statutes, trade secrecy has become the preferred method of protection. Trade secrecy is a completely separate branch of law, deriving from the common law understanding that individuals have the right to contractual protection of secrets, as well as from extremely old guild practices. Its enforcement not only further limits public knowledge of what innovations are afoot, it very seriously interferes with freedom of expression as guaranteed in the First Amendment.

Workers are forced to sign secrecy agreements that continue even when they change employer; they cannot even use ideas they themselves have worked on; courts have protected these secrets even to the point of prior restraint on the press. Sound policy would curb this sort of repressive activity as well.

Another, somewhat ironic consequence of the growing reliance on trade secrecy is that the resulting visible reduction in the number of US patents has been taken to be a sign of eroding US innovativeness, used to justify such measures as tax incentives for industrial research. Since trade secrets cannot be enumerated (if they are really secret), innovativeness is by no means so straightforward a quantity to gauge, even without taking into account to what extent it is of social benefit. Sound policies would have to be based on more detailed analysis, in part of the sort to be discussed in Chapter 11.

A last problem of the current law is that it extends the power of the state equally for protecting harmful as well as beneficial innovations. While there is no certain way to know in advance that an invention will be beneficial, it is possible to have some idea of the likely impact. Therefore, a good innovation policy would concentrate on encouraging beneficial invention by incorporating some assessment of social impact into the innovative process, and would seek to ensure that groups most likely to be affected had an early warning of what was afoot.

The goals of a democratic innovation policy can now be summarized. The policy should promote socially beneficial innovation as fast as society can in fact benefit. The innovative process should be as open and democratic as possible, recognizing that innovation is ultimately a process of change in which everyone in society participates in one way or another. There should be quick response to negative impacts. The policy should not promote inequality through concentrating wealth, nor should it encourage repressive measures.

Such a policy might be implemented in many conceivable

ways. An ultimate goal is a society in which each person can individually or collectively exercise his or her creativity protected by a guaranteed living standard and with access to a wide variety of public facilities for innovation. These facilities would include a shared compendium of technical information of all sorts, to which anyone could add.

With a good living guaranteed, rewards for invention could be non-monetary, including prizes, public recognition, and the intrinsic rewards of the activity itself.

The possible social and environmental impacts would be a topic of concern from the start of any development process, and groups to be affected could be actively involved in selecting the inventions to be developed further. All innovation would take place through direct discourse among those most likely to be affected.

The goal just described is no doubt utopian, but it does indicate the direction in which policy ought to be moving us. In that light, the proposed act would include steps that could begin immediately, preparing the ground for further changes that could be implemented gradually over a period of a decade or so.

The Intellectual Claims Act

This act would combine and revise existing intellectual property laws. Since the line between patentable and copyrightable items is already blurring, and since trade secrecy is an undesirable practice, a combined and uniform system to protect various categories of what could be called "Intellectual Claims" – excluding trademarks – would eventually help clarify the entire policy for encouraging innovation. Uniformity of treatment would have to be approached gradually, with aspects, such as the periods for which rights are granted, continuing to differ at first.

Reduction of Trade Secrecy Protection

To open the innovative process to democratic involvement and scrutiny, and to remove restraints on free expression which current legal practice entails, trade secrets should be converted to openly registered Intellectual Claims. Courts would no longer be able to enforce trade secrecy for any processes, practices, or products (with limited exceptions, as noted below) invented or discovered after this law went into effect.

For trade secrets already in existence, there would have to be an interim period during which they would still be protected, but even that protection could be limited by requiring the holders of these secrets to demonstrate good faith efforts to convert them to openly registered intellectual claims or simply to publish them.

Exceptions would be allowed for very small businesses and for a limited number of "secret formulas," such as recipes for specific food products, where these were essential to the character of the product or service for the business owning the secret. Secrets referring to specific people, (whether or not employed by the company), to current business strategies, or to other nontechnical matters would not be affected.

Prohibiting court enforcement would obviously not keep firms from privately maintaining trade secrecy agreements with their employees, so long as they did nothing illegal to enforce such agreements. Certain practices presently permitted, such as requiring lie detector tests, seem unduly invasive of employee rights and probably should be outlawed as part of this act.

Innovation Utilities (Patent Monopolies) Regulatory Commission

Intellectual property rights are granted to firms in much the same way that public utilities are granted exclusive rights to a certain business in a certain area: in both cases, the point is to encourage public benefit. Because of their monopoly status, utilities are generally closely regulated, to make sure they serve the public adequately.

In contrast, there is little regulation for corporations that have substantial monopoly rights by dint of holding large numbers of intellectual property rights, whether patents or copyrights. Yet these firms often abuse their rights in ways obviously contrary to the wider interest. At times, they use their monopolies on important innovation to garner rewards out of all proportion to their research and development costs. Also, to stem possible future competition, large firms often use intellectual property rights purely defensively, by protecting innovations they have no intention of producing.

This commission is proposed as a means of preventing such abuses. Selected to be representative of diverse interests, it would regulate all firms holding more than a certain number – say, ten – of Intellectual Claims (excluding trademarks) and deriving annual revenues of a certain amount (say, $20 million) from these claims. These firms would be defined as "innovation utilities."

The commission would gradually identify and register all such innovation utilities, beginning with the largest. Each registered firm would have to report periodically, for each Intellectual Claim held, figures such as royalties, profits, and production and marketing costs. For those items not yet being produced, it would also report on progress in development and attempts to license the product to others.

If little progress in development was reported, the firm would be asked to indicate any reasons beyond its control; a valid reason, if documented, would be that the invention is not socially beneficial.

The commission would use the submitted information to establish guidelines for prices and royalties that could be charged for each registered item. These guidelines would take into account reasonable development costs of all aspects of design, production, repair, maintenance, and use explained in the Claim at that point. (That would be an incentive to provide complete, understandable, and accurate information – see below.)

The guidelines could also be based on factors such as likely public demand as a function of price, the negative impact of unwarranted concentrations of wealth, and the goal of equality of access to technological innovations. Guidelines could be so structured that prices would tend towards a small increment over the costs of production once development costs were recovered. Eventually, these guidelines might be adjusted to influence the overall rate of innovation.

To prevent the use of intellectual property rights to stifle potential competition in an industry, the commission would also issue guidelines for adequate efforts to develop and market protected innovations.

After publication of the guidelines, the commission's enforcement powers could be phased in gradually, beginning with the most flagrant violators – those innovation utilities that fail to take reasonable actions to comply with the guidelines. Penalties could include shortening of the Claim period, termination of the Claim, and, in cases of extreme noncompliance, an order to return punitive damages to customers who paid excess prices.

A Claim not undergoing a reasonable level of development would be declared relinquished. Other firms or individuals – perhaps excluding larger firms – could then participate in a lottery for the relinquished Claim.

Licensing requirements

Eventually the innovation utility system could be augmented by the requirement of compulsory licensing of Claims at reasonable royalties. This would further limit the monopoly powers connected with incentives to innovate, while at the same time helping to assure that innovative firms would receive reasonable rewards. Further, the provision could permit communities producing largely for their own use easy access to innovations.

After a "head start" period (say, three years) to enable the original claimant to establish a position in the market, a Claim would have to be opened for licensing at royalties in accordance with guidelines set by the commission.

As the purpose of compulsory licensing would be to encourage diversity of overall supply sources, the largest innovation utilities might not be permitted to apply for these licenses. Choices between several otherwise qualified applicants might be by lot.

After an additional period for the new licensee to establish its market position, the process of opening up the Claim to new licensees could be repeated, and so on throughout the life of the Claim.

Community-based licensees

In addition to the general licenses so far described, every Claim holder would be required to grant licenses for production in and only for any specific community. Low-income communities could have license fees waived; for other communities, royalties would be based on the income and size of the community.

Developing country licensees

All holders of Intellectual Claims would be obligated to grant royalty-free licenses to developing countries for products and services to be used within that country or by other developing countries, but not for export to industrialized countries.

Monitoring Trade Secrecy Reduction

The Innovation Utilities Commission could also be responsible for monitoring the effects of trade secrecy reduction. Rather than dealing with individual violations, it would note patterns of noncompliance and cases where the law caused undue hardship. It would also look for instances in which the law failed to prevent trade secrecy from preserving monopolies.

If its findings warranted, the commission might take actions of the following kind:

- imposing price controls on goods maintained at unfairly high prices through the use of trade secrets;
- or sponsoring research to discover the technical basis of, or feasible alternatives to, trade secrets which were being used in an unfair way, and then publishing the results.

Wider innovation information dissemination and revision of patent classification system

The intellectual property system could better fulfill its Constitutional goal of encouraging socially useful invention, were records easily and freely available, easy to search through, and readily comprehensible. In many fields, the existing system of patent records could be transformed into a universally accessible (on-line database) guide to the current state of technology that would be part of the Universal Information Access Network.

A new multiple classification system could also make it easier for Intellectual Claims applicants to relate their applications to existing Claims. Provision would be made for a variety of information formats, including video, graphic simulation, etc.

Suitable incentives and regulations could ensure that Intellectual Claim applicants and holders offer as complete information as possible on production methods and problems, repair and maintenance procedures, and methods and drawbacks of using each covered invention. The information to be included could also cover the current status of development and production for each innovation.

Commission for Compatibility of Intellectual Claims with Personal Freedoms

If invasions of privacy were correctly restricted, then, as copying continues to get easier, whole categories of potentially useful innovations would become increasingly difficult to reward through the existing intellectual property laws. Our society as a whole would clearly be richer if the freest possible distribution of these innovations is permitted, provided that innovators can expect sufficient reward to continue. The following formulation should prevent invasions of privacy, while assuring incentives for creativity.

No civil or criminal prosecution involving Intellectual Claims (excluding trademarks) could depend on evidence obtained through breach of privacy or through a search warrant. The law could only be enforced on the basis of activity observable in open documents or on the open market.

The commission would identify categories of Intellectual Claims that, once the restriction described above had gone into effect, would become unenforceable as a result of newly emerging technologies (such as copiers, digital transmission of computer data, audio and video recordings, etc.). It then would attempt to find alternative ways to maintain incentives for innovation in these categories. It could take into account the cost of the innovative process, the importance of the innovation to those who benefit from it, and the number of people who benefit. Also, these new incentives should be limited, so that they provide no more wealth than would comparable, more traditional Intellectual Claims.

Items in the designated categories could continue to undergo registration in ways similar to current practices.

Where feasible, sampling methods suitable to each broad category of registered items could be used to determine the rough extent of use of each item in the registry on a periodic basis. For instance, the extent of use of different seed strains for crop plants could be estimated by genetic surveys, while the use of each book, computer program, film, etc. could be estimated by combining information on the number of requests from sample central libraries or data banks with information on total library use.

The commission would have the further duty of devising new means of encouraging creativity in fields for which the promise of intellectual property rights can no longer be an adequate stimulus. The aim would be to widen participation in the process of innovation, eliminate restrictions that now encumber the poor, make the results of innovation as widely accessible as possible, and assure an adequate degree of incentive for those innovations most valuable to society.

Reward systems that do not depend on limiting the availability of the innovations, as intellectual property laws currently do, should be emphasized. The commission could investigate ways for systematically offering innovators prizes, awards, medals, grants, stipends for future work, or income support, based on the originality, significance, influence, benefit, or popularity of their contributions.

Other possibilities include, for example, making special research facilities available to selected inventors, with comparable facilities for innovating firms. Firms might also be permitted free advertising or publicity.

The intrinsic rewards both of creative activities and of doing something of public benefit could be emphasized and enhanced by providing supplies of necessary materials, reporting to inventors of beneficial products the details of their effects, who was aided, and so forth.

Social and Environmental Impact Statement

The public, and especially groups likely to be affected, need to know as soon as possible about the implications of new inventions, preferably before full development is under-taken. Hence, each new Intellectual Claim application should include an initial social and environmental impact statement. Guidelines for these statements could be drawn up by the Social Impact Statement Commission (see Chapter 11) and the Council on Environmental Quality. These guidelines should be linked to the level of resources at the disposal of the applicant, so that efforts required to prepare such statements do not become a bar to Intellectual Claims.

Whenever a large corporation obtains a license for a claim originally registered to a less well-off person or firm, the new licensee would be required to meet the same standards for an impact statement that it would have been obliged to meet had it been the original applicant for this Claim.

The statements would be published, and also sent directly to whatever groups had expressed interest in the kind of impacts described. This could permit discussion between innovators and concerned parties about proposed uses, poss-ible modifications in the invention, and so on.

The mere requirement of these filings should increase awareness of the social and environmental impact of techno-logical innovations, and should help shift the trend of inven-tion toward greater social benefit. Also, distribution of the statements would help groups that might be adversely affected prepare to resist in advance. Likewise, groups that would benefit would have an early opportunity to encourage the spread of an innovation.

Various further steps are potentially feasible: refusing Intellectual Claims when the overall impact is negative; providing negatively affected groups special standing to sue for damages; or outlawing specific negative impact products altogether. Eventually, the period of protection of an Intel-lectual Claim could be related to the degree of social benefit that different groups in society judge the innovation to offer. Criteria could favor the most disempowered groups.

Equalizing opportunities for filing Intellectual Claims

The basis for Intellectual Property Claims has always been originality, but that can be a nebulous concept. Because of this ambiguity, large corporations involved in contests with small firms or individuals over who has legitimate rights have enormous advantages, since they can afford prolonged and complex legal battles.

Two measures could increase fairness. First, much as the state must provide an attorney for indigent criminal defendants, the wealthier side in a dispute might be required to assist its opponent in equalling its own legal expenses (with some provision to prevent the less wealthy side from initiating utterly frivolous disputes).

Second, the Intellectual Claims Office could establish, and advertise the existence of, a national network of regional assistance centers to help prepare applications, along with social and environmental impact statements. The office could charge for this service on a sliding scale, with the fees being set at zero for low-income private individuals or for businesses below some minimum size.

Employee rights

Although most intellectual property that is remunerative is assigned to corporations, patents are only issued in the name of individuals, with most employees required by contract to assign patents to their employers. Currently, employees who make inventions in the course of their employment have little or no right to determine the fate of their inventions. Employers are free to ignore employee inventions, and do not even have to apply for patents in order to prevent their employees from doing so independently. Finally, employees often receive little or no reward for patents in their names.

This act would prescribe a definite period, such as six months, from the date an employee informed his or her superiors of an invention for the employer to pursue filing an Intellectual Claims application in the employer's name. Failing this, the employer would relinquish any claims to the invention, contractual or otherwise. The employee could then file her or his own application, as a private party. If the employer did exercise the rights to file, the employee would be guaranteed a fraction of all royalty payments received for licenses, and a similar fraction of all profits resulting from sales of the invented article or process.

The employer could refuse to develop the invention because of an overall negative social or environmental impact, but would have to substantiate this judgment with suitable evidence. Otherwise, if the employer failed to utilize or develop the invention within a specified period, such as six months, the employee would have the right to take over the Claim from the employer and pursue other avenues to develop it.

Registering claims upon publication

Currently there is a waiting period between the time a patent is applied for and the award of a patent, even though the period often does not afford time for careful examination of the merits of the Claim. Publication of the idea of the invention more than a year prior to application can invalidate a Claim.

For four significant reasons, the law should eventually permit registration of a Claim (equivalent to granting a patent) upon publication of the essential idea of the invention. First, as the innovative process is basically social, exchanges of ideas at the earliest possible stage will help promote better products and services, more fully in tune with public needs. Second, a principle of registration of Claims on publication will counter current trends towards keeping discoveries secret until a patent application is filed, which increasingly is turning scientific discovery into the private wealth of corporations. Third, for an action such as invention, the only unimpeachable way to establish Claims to priority is to act in public view. Finally, it has been customary for copyright to be granted upon publication, and uniformity among the different categories of Intellectual Claim would make the law easier to understand and utilize.

Objections to patent on publication have been that claims would be awarded to rediscoveries, to trivial modifications of existing inventions, or to vague and worthless inventions. These would presumably clog the record of patents, making valuable ones more difficult to locate. But a beneficial and useful rediscovery is just as much to be encouraged as a new invention.

Attempts to patent trivial modifications would not be a real problem, since only the Claim for the modification would be independent; if it were deemed unimportant, it would neither profit the inventor, nor in itself harm any one else. The only concern would be the growth of records of Claims and the greater difficulty in wading through them. A suitable classification system and ready availability of Intellectual Claims information, coupled with requirements of clear and complete filings, would make vague or redundant claims easy to ignore.

Filing a Claim could amount to publishing a description in any periodical with some minimum circulation, and forwarding a copy to the Intellectual Claims Office. As an alternative to publication, the office could make available its computerized information system (tied to the Universal Access Information Network).

Other government influences on the climate for private sector innovation

Besides the intellectual property laws, government influences on innovation include tax incentives, technical information, research applicable to private sector use, opportunities to take over government-held patents, measurement standards, and foreign trade regulations.

In general, these activities are of greatest use to large

corporations and do not particularly favor socially beneficial inventions. These programs could be altered along lines parallel to what has already been proposed.

Consider, for example, tax credits, which amount to a loosely controlled form of government expenditure. Tax expenditures of any kind could be a valid means for government resources to be used to encourage particular categories of private actions (e.g., charitable donations), while allowing a wide leeway in interpretation and in approach. In practice, such expenditures often turn out to give greater leverage to the well-off, so that the actions sponsored are likely to reflect the interests and values of this group. Moreover, unlike other forms of government expenditure, there is no regular review. Neither of these drawbacks appears insuperable, however. Tax credits – as opposed to deductions – could redistribute power, and regular reviews of what benefits result from the policies could be instituted and could then lead to changes.

At present, research tax credits, investment credits, and other tax expenditures influencing technological innovation may help large corporations maintain an advantage over smaller competitors, or may be used for socially undesirable activities, such as automating enjoyable, well-paying jobs out of existence. But the Innovation Utilities Regulatory Commission could help ensure fairness and public accountability. Likewise, procedures described above under Social and Environmental Impact Statements could apply to tax expenditures as well as to Intellectual Claims.

The research and development programs outlined in Chapters 8 and 9 and the agencies for reviewing social impact to be described in Chapter 11 would provide a guide for modifying all existing government programs that significantly affect private sector innovation and the innovative activities of individuals, communities, and smaller corporations. That would involve, for example, increasing efforts to present technical information in understandable forms, a change of emphasis in such areas as standards of measurement, widening the range of groups participating in planning policies and projects, and generally promoting a more open and accessible technology community.

Finally, despite the convenience of placing the issue of the climate for private sector innovation in a chapter of its own, it should be recalled that private sector innovation is in fact directly influenced by government research, development, and procurement policies, so that the proposals for new programs in the two preceding chapters, added to the proposal for reducing defense spending to be discussed in Chapter 12, would directly – and desirably – affect private sector technology as well.

11 Technology rights, review, and regulation

The preceding three chapters have dealt with government programs to stimulate the emergence of desirable innovations, both public and private. This chapter concentrates on government and citizen responses to the existence of new technologies.

One of the commonly accepted roles of government is attempting to alert people to phenomena, such as the weather, that are likely to affect them. If fine weather is approaching, a timely notice permits planting crops or planning a picnic; storm warnings provide time to protect crops or seek shelter. If the damage of the storm is too great, government routinely acts for society as a whole in providing the victims with recompense and assistance in rebuilding. And if someone is without shelter, then government has some obligation to step in and see that it is provided. This function involves attempting both long- and short-range forecasts.

Unlike the weather, technology is under human control. That does not mean, obviously, that each and every person is in a position to control the technology that can affect her or his life, but it should be the aim of a technology policy to move towards that condition. As is true for the weather, it makes sense for the government to be genuinely attempting predictions, both long- and short-range, of immediate impacts of particular technologies and of overall impacts.

But just as storm warnings are of most help if there is

shelter available, knowledge of likely effects of new technologies would be of most use if there were ways to act on that knowledge. That holds true even of the technologies that would emerge from the social goals-directed development described in Chapter 8. What does it take to be able to protect oneself from, or to avail oneself of, a technology? It requires not physical shelters but legal ones – rights, broadly speaking. When it is possible to define rights so that each person is able to make use of them for the protections and opportunities they are intended to offer, legal rights are sufficient.

In addition, in an area such as technology, to exist at all, rights often have to be exercised collectively. Even that does not suffice in all cases; when groups are not well enough defined to exercise rights, they need government regulation to protect them.

Social Impact Statements Commission

Because technologies spread so rapidly, even a single invention can have a very broad impact in a very short time. Fortunately, an even faster spread of information about the potential impact is possible if preparing such information can become a standard part of the development process. That would be the purpose of the Social Impact Statements. In order to be functional, the statements would have to be fairly simple. The more complex analysis of the potential impact of an innovation becomes, the more chance there is for deliberate or accidental misinformation, confusion, and obfuscation. Moreover, the easier the statements are to prepare, the less burdensome they would be and the less the attraction of evading the obligation. The Social Impact Statements Commission would devise means of providing statements that meet these requirements as well as possible: short, usable statements that nonetheless pinpoint significant impacts.

At first glance, the task might seem impossible. Gauging

possible impacts in advance may seem difficult enough without being reduced to a simple or standard process. However, the system would not have to approach perfection to be a huge improvement in comparison with the current provision of no system at all. In addition, it would not be a requirement of Social Impact Statements to reveal the sorts of broad change that can result from the interaction of many new technologies; these would fall under the purview of the commission to be described in the next section. With this understood, a reasonable questionnaire approach is possible, since the vast majority of innovations depart from previously existing technology in only limited ways. Those departures themselves, furthermore, usually resemble in character departures that have already occurred in other kinds of technologies. For instance, a new production process in the office furniture industry might resemble a similar innovation in another industry, such as the household appliance, automobile, aircraft, or building materials industry. Many of the work-related effects might be immediately comparable in the two industries. The changes in the final products emerging from the new process would also probably be guessable by analogy and would be a clue to the kinds of changes in offices – perhaps affecting workers or clients – the innovation might lead to. Similarly, any kind of device permitting new forms of home entertainment raises questions that television also has raised – for instance, about how they will complicate the abilities of parents to maintain some control over the cultural environment of their young children.

If an innovation does not seem closely analogous to previous innovations, that is an indication that its possible impact merits special reflection. Even so, that would not necessarily imply guessing blindly. Strikingly new departures, such as the automobile, the phonograph, nuclear weapons, or the personal computer, have had major consequences that could have been predicted on the basis of generalizations from the past, taking into account the particular scales – e.g., of size, affordability, or speed – in which each differed most sharply from predecessor technologies.

In addition, just as most inventions can be related to prior inventions, many of the important social impacts also can be described very well in terms of just a few categories, such as who would be likely to gain in wealth, or in power, and who to lose, how contacts within the community would be affected, the size of the community that could easily be influenced by óne user of the invention, how the interest and flexibility of work and levels of employment would be directly affected, and so on. The commission and its staff would refine these insights and, starting with one standard questionnaire, would create more and more specialized questionnaires for different kinds of innovations. The efficacy of the questionnaires could be tested by using earlier innovations as examples. Continual refinement of the process could lead to more and more useful information without unduly burdening would-be innovators. As soon as feasible, the commission could move from preprinted forms to online computerized questionnaires, which would permit still more appropriate sets of questions, depending on the nature of the innovation. This system would allow the commission staff to assist in cases in which the innovator was uncertain how to answer. Instances for which unusual impacts might result could probably be picked out and subjected to more careful scrutiny by the staff.

One reason that Social Impact Statements can be quite different from Environmental Impact Statements is that the environment cannot speak for itself, while social groups do have voices if they are allowed to be heard. The Social Impact Statement process must permit groups likely to be affected by an innovation to speak in time; it need not predetermine the thrust of their response. By the same token, preparing and filing the statements would only be the first step in a social dialogue. (The process of dialogue would also have to include a voice for the potentially affected groups in determining the categories and question-naires themselves.) After that, the statements would have to reach the attention of the likely repondents. Chapter 10 described a mechanism of doing this in the case of innovations covered by Intellectual Claims.

The chance to obtain an Intellectual Claims registration would be the major incentive for filing a careful Social Impact Statement. But, as for now, not every significant innovation is protected as intellectual property, and even the revised definition of Intellectual Claims might not induce every innovator to file. As long as exceptions are infrequent, and follow no special pattern, they would be little cause for worry. Someone, and it might reasonably be this commission, would have to survey innovations periodically to gauge the importance of the exceptions. If they became too numerous, or if significant patterns of non-registration connected with possible negative impacts were to show up, then the requirements for filing Social Impact Statements would have to be widened, and with them might have to come additional avenues for public debate.

The Commission on the Overall Impact of Technology

Existing policies, as well as the alternatives described in this book, both assume a fairly high level of technological change. If our society as a whole is to maintain any sense of where it is going, the synergistic impacts of all these technologies taken together, and not just individual ones, need attention. Such attention should not be occasional, but continuous. While there are now a variety of forums in which these broad changes are contemplated from time to time, ranging from journalistic accounts to academic studies to congressional hearings, there is no single body, accountable to society as a whole, with adequate resources, that has as its chief purpose drawing together, on a regular basis, a clear picture of this complex whole.

This commission would be such a body. It would be responsible for issuing a report to Congress and the public at, say, five-year intervals on its current assessment of the overall environment of technological change and the impacts upon society. It could, at its discretion, issue emergency reports. It would have the duty of proposing alterations in the overall direction of technology policy, of suggesting, if necessary, limitations on the rate of change, and of suggesting ways and areas in which to discourage innovation, as well as to encourage it. Members of the commission would have to be given considerable independence, so that they could withstand such forces as the business community and other government agencies. Their own backgrounds would have to be diverse enough not to predetermine their conclusions.

With the same sort of continuity and singleness of purpose as the full commission, subcommissions could focus on the impacts of technology on work, on international relations, on community life, etc. Additional subcommissions might concentrate on aspects of technology that are particularly difficult to gauge and likely to be especially important to watch, such as the impact over time of computerization or of highly complex technologies.

Rights of workers, consumers, and communities

For humane technologies not only to be developed but actually used, those who would benefit from them have to have power either directly to put them into effect, or to argue strongly for them. In addition, workers, consumers, women, ethnic minorities, and local communities need means to protect themselves from socially as well as environmentally or physically harmful technologies. In most general terms, people must have the right to be involved in technological decisions that will affect them, and therefore a right to be informed in a timely manner of technological alternatives, and some legal means to enforce those rights.

What such rights mean would vary sharply in different circumstances. As workers, people currently have quite limited rights to influence technological decisions affecting their jobs. In part, this is a consequence of existing labor laws. As consumers, the same people, especially if they have money, have something of a free choice, some chance to know of drawbacks, but rarely much direct influence over what products are to be available; also, high technology products often become useless due to changes made by their manufacturers, including simply going out of business. Communities that are political entities, such as incorporated cities or towns, have legislative powers that can regulate some technologies; other communities have much less direct power. The poor, minority groups, or women each have no common voice, and thus have little opportunity to act cohesively, in addition to their lack of legal standing to do so in many instances. And of course, any of these difficulties transcend the area of technology. Hence, meaningful provision of the general rights described above must be arrived at piecemeal, along with other extensions of democracy.

There are a variety of ways in which rights may be enforced in American tradition: enforcement power can rest with a regulatory agency, through injunctive relief, or through the right to collect damages after the fact; rights to negotiate contractual provisions dealing with specified matters can be guaranteed by statute; infringing certain rights can be made a crime; local communities themselves have the right of legislation over certain activities within their jurisdictions; finally, of course, there are means such as boycotts, disinvestment campaigns, and other actions based on negative publicity. Some of these means, especially the last, are clearly already available. Additional enforcement procedures should be selected so as to ensure that the new rights can be exercised easily and effectively in the appropriate situations. For such a broad area as technology, no one enforcement means would work well in all cases. In some settings, especially workplaces, there is a possibility for effective negotiation between the two parties (workers and management). However, negotiation alone cannot work when there is no effective, well informed, collective

bargaining group that is in a position to foresee the consequences of management actions and act with strength; in
these circumstances, some *ex post facto* means to obtain
damages would be justified. That is the reason for the
malpractice provisions below. In other cases, a technology
has too widespread an impact, or involves too many
different groups for direct negotiations to work. In such
cases, direct government regulation may be appropriate.
Finally, a technology may be of great social benefit but still
have harmful consequences for particular groups. In this
case, compensation of some sort ought to be devised at the
earliest stage possible. Each of these mechanisms should be
aimed towards overall social improvement, if possible by
encouraging cooperative and imaginative attempts to find
means – which might well be additional technological means
– to change existing conditions in ways satisfactory to all
concerned.

The next section of the chapter focusses on regulation
and compensation. The agencies proposed in Chapters 8
and 9, along with Social Impact Statements, could promote
a number of these rights, by generally assuring sources of
information, by direct contact with groups they are intended
to benefit, and by providing improved means for communication and coordination for groups such as women and
ethnic minorities. The following indicates how more
narrowly focussed legal rights could be specified.

Workers' rights

The work situation is the one where responsibility for new
technology is easiest to pinpoint, and where many of the
impacts are very clear. Workers' rights might be expanded
to include the following:

All employers would have the obligation to inform their
employees of every contemplated technological change that
would affect the nature of particular jobs, the number of
jobs, overall productivity in an industry, means of super-

vision, flexibility of workers to determine their own pace, or any other change known to affect aspects of employment. Employees then would have the right to obtain full information about the contemplated change, to be kept informed of subsequent planning or alterations in proposals; workers likely to be affected could meet with other workers for reasonable periods of time at work or elsewhere, during designated working hours, to discuss the effects of these changes. They could also go over the plans with outside consultants of their own choosing. Acting together or individually, they could propose modifications and negotiate with the employer on the form the actual change, if any, would take. These negotiations would not be limited as to subject, including disposition of resulting profits, reassignment of workers, deciding who would be able to use the new technology, and for what, examining further alternatives, etc. Before any orders for new equipment or tools could be placed, however, the workers to be affected would have the right to vote to accept or reject the order. In the event they rejected it, negotiations could continue, possibly to be terminated eventually in arbitration.

Workers also would have the right to urge their employer to undertake a technological change, explaining the benefits they believe the change would confer. The employer could not reject an affordable change if it would offer improvements in working conditions, flexibility, or comfort of work, improve skills of the workers, improve the product or productivity, increase the range of products that could be produced, or offer another clear benefit. Disagreements about the benefits or affordability of the change could be arbitrated, under the assumption that a fair fraction of profits from that particular work unit could legitimately be used for such purposes. Other workers to be affected would have the right to intervene on the same basis as if the employer had initiated the change.

All workers working for a particular employer, regardless of location, would be covered. If several different places were involved, the employer would have to provide adequate means for all affected workers to communicate, in privacy, with each other. Workers would be assumed to be affected by a technological change not involving them directly, if, as a result, their own work unit would be likely to become outmoded or otherwise declared redundant and closed. Since these results might be the consequences of decisions taken by employers other than their own, rights to know of and intervene in such decisions must also be considered.

Malpractice

Employers would be understood to have a duty to workers to ensure them the chance to improve their skills, learn new skills, improve the conditions of their work and their value to the larger community, continue to be able to work in the same place, to have jobs and to be employable on the outside job market. Failure to take reasonable actions to meet these duties would constitute management malpractice. Each worker would retain the right to sue for malpractice the employing corporation and/or any managers supervising that particular worker directly or indirectly. The damages could comprise lost wages, loss of improved conditions, and related losses, plus punitive damages if the malpractice could be shown to result either from deliberate planning or gross negligence. (Unlike claims of medical malpractice, which sometimes stray into punishment for what amounts to bad luck or failure to be superhuman, this law could be effective while being limited to patterns of actions that demonstrate prolonged misconduct or unwarranted incompetence. For fairness, it could also include a form of graduated responsibility, depending on the size and power of the employing business, factors which would affect its ability to have adequate information and opportunities for acting.)

Consumer Rights

These rights are already partially assured through warranty obligations and existing consumer protection legislation. They need augmenting as technologies become more complex and as corporations have an easier time changing identity. Although only barely relating to technology, the 1985 decision of the Coca Cola Company to change the flavor it had been advertising for years as the "real thing" demonstrates the remarkable power and arrogance of larger corporations towards consumers. In that, as in many other cases, the issue for the company was not any unavoidable loss, but simply a declining market share. The rights proposed would help combat such blatant unilateral decisions.

These rights would parallel workers' rights in several respects. Groups of consumers who had previously been served by a company would have the right to require that that company keep selling them a product it previously made available, or that it produce some additional item that it should have the capacity to make, that it modify a product along certain lines, etc. As a group, they should have rights comparable to workers' rights to communicate with each other, to negotiate with the corporation, to have access to relevant information from the corporation, etc. The corporation could avoid meeting the requirement if it could show that its reasonable costs would lead it to lose money on the particular transaction in question, or that a suitable item was available from another source.

A single consumer or a group could sue a corporation for malpractice if it had failed to devote reasonable efforts to improving the products it sold, if it had failed to inform a prospective consumer of the drawbacks of a product or of well-developed plans to withdraw a product from the market or replace it with a superior one that the consumer might have preferred to wait for.

An additional protection is needed for consumers who buy complex technological products from new firms that may quickly disappear. Some form of government-mandated warranty insurance could suffice. To qualify for the insurance, the firm would have to provide detailed repair plans, together with some plans for supplying spare parts. The warranty should also be extended to cover failure of the design to meet reasonable expectations.

Community rights

Social groups – for example, women or minority groups – as well as specific communities can suffer from the failure of a certain technology to be developed, or, once developed, to be produced, as much as from what is developed and produced instead. In most instances, dealing with such broad issues might best be left to regulation, as well as to non-legal means of affecting corporate actions. But there may be situations in which direct exercise of rights would be appropriate. One example would be a particular firm's refusal – without adequate reasons – to develop versions of its products in forms easily usable by average women. In such cases, the right of filing a class action suit to compel the change, as a last resort, could back up efforts to negotiate. Likewise, the right of filing a malpractice suit after the fact might help prevent a firm from undertaking an unwarrantedly male-oriented development process that could be expected to have a substantial negative effect on certain women. Rights such as this would reinforce the desirability of an early filing of a Social Impact Statement. If a firm anticipated that its actions might cause future harms to identifiable groups, the existence of these rights would prod it to negotiate a more mutually acceptable direction.

Commission for Emergency Regulation of New Technology

New technologies traditionally have spawned new regulatory agencies: the Interstate Commerce Commission, the

Federal Aviation Agency, the Nuclear Regulatory Commission, the Federal Communications Commission, to name a few. The chief identifiable constituency of most of these agencies is the industry they are supposed to regulate; the public interest can get lost in a myriad of detailed decisions comprehensible only to specialists. Because too numerous regulations can become unworkable, the approach in this book has been to attempt to keep direct regulation to a minimum, and to replace it with positive government measures to assure that new technologies will promote desirable goals. For instance, the Agency for the Mass Distribution of New Technologies would help give everyone an early opportunity to make use of technologies that otherwise would have given increased advantage only to the rich. The alternative would have been to regulate new technologies, slowing their introduction if they were to offer such limited benefits. While regulation can be kept to a minimum, the power of new technologies is such that regulation of the new cannot simply be abandoned, at least as a last resort.

The previous section described a range of legal rights that would be most useful for organized groups or individuals who could pinpoint both the nature of a benefit or harm and the specific firm (or other organization) responsible. Regulation is needed in cases in which those rights would not avail, because the group in question either was not sufficiently well organized, lacked the means to pursue litigation, or could not pinpoint any one responsible firm or group of firms, or any narrowly defined source of harm. Consider, for example, the profusion of new technologies that make it easy for firms to contract out work to people working at home. The harm there is that working at home makes adequate health, safety, and minimum wage standards difficult to enforce. The workers, often single mothers with small children, are isolated from one another, have little opportunity to organize collectively, and are subject not only to the stress of performing two tasks – childcare and the specific job – at once, but are left with little time for normal social contacts. For some of these reasons, home work has long been outlawed, although such laws have never been easy to enforce, and, at certain times and places, have been widely ignored. At present, when a new technology makes ignoring this law even easier, by permitting office work at home via telecommunications, a regulatory agency that could affect the course of this development or propose compensatory mechanisms would be of great value.

Regulation is normally retrospective; it is only after technologies have been introduced, and the damages from lack of regulation have become evident, that Congress has mandated new regulatory agencies. Even then, it is often the case that a particular agency is forced to delay – often for years – implementing a regulation that would stop a certain undesirable practice. Such delays themselves often stem from the fact that a new technology is already in such wide use that powerful interests oppose any regulation; delay then entrenches bad practices further. Whenever there are reasonable means to guess at damages in advance, prospective regulation makes more sense, and may be much more equitable for all concerned than after-the-fact procedures.

Here are some situations – obviously still needing more precise definition – when advance regulation would probably be warranted:

- when the effects on the poor – whether all or only scattered individuals – or other very disempowered groups would be adverse and significant;
- when local or firm by firm actions to enforce rights would not likely work because each firm could plead that overall levels of competition in an industry determine how it must act to survive;
- when new technology would lead to dangerous concentrations of power;
- when the overall rate of technological change is simply far too high for society as a whole to absorb without serious disorders;

- when a new technology might provoke grave international repercussions;
- when cultural effects would be rapid and unpredictable, or predictably quite bad.

The commission would obtain its evidence for potential damages through the Social Impact Statements, from reports of the Commission on the Overall Impact of Technology, or through its own assessment of petitions addressed to it from any source.

The regulations promulgated should at most amount to temporary bans until a way out of the difficulty can be devised, unless the new technology had almost no redeeming features – e.g., parts permitting easy assembly of operable machine guns. If, in the case of some particular technology, more permanent regulations seemed necessary, this commission would have to propose to Congress a new agency for that specific regulatory task. This proviso would prevent one agency's accumulating a jumble of different responsibilities that would make public awareness of its activities difficult. If regulation does have its undesirable side, keeping it as clearly as possible in the public eye is one way to mitigate the evil.

The commission could either propose ways to modify the technology so that harms would be minimized, propose ways for the groups who would benefit to compensate those who would be harmed, or, in a case where society as a whole would clearly benefit, order (or propose) direct compensation in monetary awards or in the form of other appropriate government action to the specific groups and individuals harmed. It would attempt to choose the course most favorable to those harmed as well as to the country as a whole.

In the example of home work, discussed above, the commission might require that the communications system used be modified to permit direct links among workers so that they could organize or easily and privately make complaints to government inspectors, along with reliable recording of hours worked and wages paid, in ways open to government and union inspection, to prevent underpaying. Additionally, it might require the industries involved to provide funds for good childcare facilities near workers' homes.

12 A model "Peaceful Global Technological Cooperation Program"

The capacity to develop new technologies is now wide-spread, existing in many countries, even in a number, such as India and Brazil, in which large parts of the population still live very much on the margins of technological society. Technology is also universalizable, we can expect what we do to spread, and we may very well find ourselves adopting what was developed elsewhere. Existing technologies already have helped promote a world of interdependence in which no one nation is sovereign over its own fate. Trans-national corporations, by their very nature, are able to evade most efforts to exert any sort of sovereign control, often by the simple threat (or act) of closing up shop and moving elsewhere. They are in an obvious position to make technological choices affecting any of the countries they work in. Since today virtually every corporation above some tiny minimum size could operate globally if it so chose, any attempt to influence business decisions – such as those described in the two previous chapters – would be weakened severely if the global connections are ignored.

World-wide immigration patterns alone would eventually undercut any program that does not address their roots. If the only way to maintain a standard of living above the world average is to be cruel to would-be immigrants, we could hardly maintain a humane or democratic society within our borders. Beyond all this, the global level of armaments continues to rise, and if arms everywhere continue to increase in sophistication and range, every country will eventually be threatened by all the others. None of this is cause to abandon hope, but it does impel a strong international dimension to the overall program – based, it should be clear, on cooperation rather than coercion. What that might entail is suggested below.

Agency for Global Technological Cooperation

At present, the dominant attitude of existing American technology policies with respect to international relations is to emphasize competitiveness as a necessary response to the fact that heightened global trade and military development have reduced, and threaten to reduce still further, our position as world leader. Supporters of that viewpoint could be expected to challenge many of the proposals in the four preceding chapters, at least on the ground that the realities of international trade do not permit us to move in any direction other than higher productivity.

That argument would be mistaken on at least four counts: first, as seen in Chapter 2, the universalizability of technology means that our efforts to promote our own competitiveness will simply intensify the competition against us – which is particularly relevant in a period when there are many potential competitors, with far lower wage rates, who could rapidly adopt our technologies; second, quite a few of the programs described in Chapters 8 through 11 would help us increase our level of internal economic self-reliance, and so lessen our dependence on international trade and vulnerability to competitive pressures; third, if we adopt these other proposals, interest in them will certainly build elsewhere; finally, as the premier technological innovator in the world, the US, if it uses its position wisely, is far more capable of resisting external technological pressures than any other country.

The Agency for Global Technological Cooperation would work to try to refocus international technological efforts along lines similar to those proposed for this country, including promoting economic equality, raising the maximum achievable level of community economic independence, free exchange of technological information. In addition there would have to be a coordinated effort to aid the poorer countries in raising their living standards to levels of satisfaction comparable to our own by some agreed date, while undercutting their social and cultural patterns as little as possible and encouraging protection for natural environments.

The agency would seek noncoercive ways to encourage each of the other advanced countries to join in each of the specific research and development proposals discussed in prior chapters. It would further seek means to promote technologies permitting heightened international communication among members of groups such as women, workers, and the poor, who share common interests across borders.

To overcome the negative effects of technological competition, the agency would seek agreements on limiting productivity increases to situations in which the countries likely to be affected already had full employment and roughly equal levels of guaranteed income. It would also seek agreements to devote excess industrial capacity to aiding Third World countries in reaching the goals indicated above. As was emphasized in Chapter 3, when Third World countries receive development aid in the form of loans or investment to make repayments, they are forced into international trade at a disadvantage and must depend on low wages to enter markets; that impoverishes their people and drives down worldwide wage levels. To prevent this, the proposed aid would be in the form of outright grants and would support development to meet internal needs. It would include means to make use of technological information supplied freely by cooperating advanced countries.

The Agency would attempt to obtain agreements among major advanced countries to set a target date in the first third of the twenty-first century, by which time they would have provided sufficient help for participating Third World countries to reach living standards that would offer satisfactions of the same level as in advanced countries. (One way it might be possible to tell that people were equally satisfied, despite different cultures, etc., would be when net immigration rates across international boundaries reached zero, i.e., with as many people going as coming; this would clearly be meaningful only in the absence of restrictions on such travel.)

Peaceful Technology

Research or development related to weapons and other military purposes is counterproductive for at least six reasons: first, it violates the "categorical imperative" of technology – not to develop what you would not want an opponent to develop – in effect pointing weapons against oneself; second, new weapons, even if not matched by an opponent, increase the complexity, uncertainty, and danger of international relations; third, new weapons themselves help incite a martial attitude and bellicose spirit; fourth, in a finite world, new weapons keep moving us towards more instantaneous annihilation, placing everyone more and more under imminent threat of death; fifth, military technology skews all technology, thus moving society itself towards militaristic values; sixth, producing ever new weapons also explicitly increases the possibility that our society and others could directly be taken over by militarists.

Military research and development should therefore be ended as soon as possible. There are two widely recognized obstacles: the apparent difficulty of obtaining verifiable international agreements to stop research, and the economic impact of eliminating research, development, and procurement of new weapons systems.

An approach to achieving mutual reductions in weapons research, development, and deployment

Realistically, at present, the only potential enemy worth considering in relation to the supposed need to keep devising new weapons is the Soviet Union. The existence of mutual enmity is attested to by the sheer number of missiles the two countries have aimed at one another; yet, ironically, the magnitude of existing destructive forces is so great that it allows considerable leeway in taking unilateral risks for disarmament. In particular, even a several year unilateral halt in research and development of new weapons would pose no danger to our security. Existing methods of detection (and espionage) would make it easy for the Soviets to verify that such a halt was very nearly complete. Given the danger of the arms race, as well as its expense, they would have enormous incentives to copy all the verifiable aspects of the halt rapidly.

These verifiable aspects currently would include a complete halt in: missile tests; detectable underground bomb tests, which at the least would cover all but the very smallest nuclear weapon; the flight testing of new airplanes; sea testing of new submarines; and probably much else. Other means to verify a research and development halt would include, for example: allowing inspection of all satellites before launch; allowing inspection of large industrial or research facilities on some regular basis; opening scientific conferences to visitors; allowing free exchange of scientists, with visiting positions at every significant laboratory; permitting examination of large-scale computer operations; and limiting secrecy.

A major downward reduction in the development of new weapons would be so worthwhile in itself that it would outweigh any risks of small-scale research projects going undetected. However, to further minimize that risk, one form of research worth continuing and expanding would be devising new ways to assure that surprise weapons were not being prepared. Soviet cooperation could be rewarded by further measures at openness, reducing hostility, and economic cooperation.

Claims that unilateral stoppages are dangerous ignore that decisions to proceed with research are equally unilateral and, as history has repeatedly demonstrated, decidedly dangerous. A unilateral research moratorium would offer several obvious advantages relative to a negotiated agreement. Work would stop sooner; fewer weapons would therefore end up being developed. As soon as the other side responded, further unilateral acts could follow in quick succession; negotiations towards a treaty, by contrast, can take years to arrive at a complete package of agreements. Public pressure to maintain and increase the moratorium would be far more effective than attempting to monitor the activities of a small group of negotiators operating in secret, each likely to have ties to vested interests that could be hurt by an end to research and development.

An approach to the economic consequences of a military technology halt

Military spending generates demand that has become an apparently essential prop to our economy. With heightened internationalization of the market, the fact that such spending is very largely domestic makes it even more difficult to stop, since military spending can fairly successfully be limited to domestic channels, despite avowed support for free trade. (The need for assured supplies in case of war, for example, apparently legitimates procuring weapons components internally.) Additionally, military sales are a byproduct of our weapons development and an area of trade surplus.

Yet we and the world as a whole would clearly benefit if the societal resources now devoted to defense could be redirected towards more beneficial ends. What has always made that kind of redirection politically difficult is that in most economic areas public spending is in some sort of competition with a portion of the private sector. For instance, public parks compete with private amusement parks; mass transit competes with automobile sales and parking garages; public schools compete with private schools and tutoring businesses. Even welfare spending is seen to compete with the work ethic. A long-term answer to replacing military spending would therefore probably have to include such measures as a restructuring of the international economy to lessen the negative effects of competition, increasing the intrinsic interest of work – so that less reliance need be placed on coercing work by the threat of poverty – and better balancing the distribution of incomes in the country. The proposals elsewhere in Part Two would aid this transformation.

If military spending is to decline as soon and as sharply as it ought to, shorter range, partial, stopgap methods would be needed also. The proposals of Chapters 8 and 9, along with increased aid to the poor both in the US and in the Third World could probably take up much or all of the slack. In considering these and other possible replacements for military spending, we might spare ourselves much fruitless political wrangling if we bear in mind that the primary economic importance of defense is nothing other than the need for waste in the international system of which we are part: without waste our mechanisms of production outrun our mechanisms of distribution, and supply exceeds demand. Hence, while wastefulness is not a virtue, calling a program that would temporarily substitute for military spending wasteful would not in itself be a cogent argument against it. The question cannot be whether it is possible in principle to find a replacement – surely we are that imaginative as a nation – but only what the best choices would be.

As emphasized in Chapter 8, programs that do replace military spending should not involve continuing the same institutions, be they laboratories, testing facilities, or factories, without major changes in organization and personnel. Otherwise, it would be too easy to reconstitute military programs. In every region of the country, a program of conversion from military to civilian spending should include a diversity of agencies, both federal and local, with a variety of social goals.

Further, in funding replacements for military spending, it should also be a national aim to contribute most to areas of highest overall unemployment, and not necessarily to assure new jobs at equivalent levels to the same individuals who had previously been well provided for by defense contracts. (Obviously, hardship for former military workers should be averted.)

An approach to world order

The US military system, it is sometimes claimed, by both supporters and critics, has helped maintain an international political and economic order which has favored America, or at least some Americans. In the period from World War II to about 1965, that claim might have had some validity, but it would be difficult to correlate what limited international order there is now with the presence of American force. The psychological trauma of continually being thirty minutes away from a nuclear cataclysm alone is ample cause for doubt.

In addition, the attempt to find ways to "project" military power has often distracted our politicians from any rational effort to promote even enlightened American self-interest, much less a disinterested world order. US actions in many parts of the world – Southeast Asia, the Middle East, Turkey and Greece, the Horn of Africa, South Africa, Chile, Central America, Western Europe, and New Zealand – attest to that basic distraction. Helping to promote and maintain a genuine order based on peace, mutual respect, and human rights would involve quite different activities, and would be a goal to which a nation as large and wealthy as ours could certainly contribute much.

Multilateral agreements comparable to the one with the Soviet Union just discussed could help slow arms races and reduce tensions worldwide. They could take into account conventional as well as nuclear weapons. One aspect would be ending the practice of trade restrictions on technologies of supposed military value, in concert with other measures to promote eased tensions.

An additional worthwhile replacement for military research would be a much enlarged effort to understand the causes of international tensions and wars, and to find ways through diplomacy to help countries steer toward just internal and external peace. Wars are not only terrible in themselves, but have helped to promote and obscure almost every form of evil, from genocide to famine, that this century has seen. Wars are not random events; they are fought for quite definite reasons. Therefore, using modern information gathering capacities it should be possible to foresee likely wars in or among the 150-odd countries soon enough to find non-military ways of settling the underlying grievances. Neither the US nor anyone else seems to have made the relatively modest effort – involving perhaps a few thousand workers – to do this adequately. It should be a priority.

13 How would it work?

As any reasonable attempt to revamp technology policy must be, what has been outlined in the last few chapters is a rather complicated system. Could it work? What would the new technologies be like? How about the new sciences? Could all these programs lead anywhere, or are they just so much fantasy? Can we really improve our economic order? How much would this program cost? These are among the many questions the presentation might have provoked. A fuller sense of what a reinvented technology would entail should emerge from the answers.

– Why is it necessary that we reinvent technology?
Technology's current direction, linked with prevailing economic and political forces, threatens calamitous changes, if not for everyone, then for most people in the world. The dangers include war, unemployment, weakening ties to others, personal insecurity and social instability, a sharply lessened sense of one's role in the world, all culminating in loss of personal power over one's own circumstances. The speed of technological development means all these trends can suddenly assume crisis proportion. At the same time, many technologies now available could help move society in far more promising directions.

– What kind of world would a reinvented technology help bring about?
A world in which the resources and capacities of ordinary people and communities would be greatly augmented, in which connections among communities would also be increased, and in which at least the sharpest and most evident sources of tension and mistrust between communities as well as outside them would systematically be reduced.

Arriving at the new technologies

– What would the new technologies be like?
No doubt, it would be far easier to contemplate the policies put forward here, were it possible to describe the new technologies in advance. But the point of the program is to stimulate an inventiveness along lines that have been little encouraged up to now. This would be the work of hundreds of thousands of people, and any attempt to guess at the details would have to concentrate on what is known already, that is, on the very parts that would be the most unsurprising aspects of the whole open-ended undertaking. At first, the new technologies, taken individually, would resemble current ones. What would be most different would be the sum total of changes, the directions not being pursued, as much as the ones being pursued.

The easiest question to answer, then, is what the new technologies would not be. They obviously would not be weapons-related. That in itself would lead to less emphasis on technologies of enormous scale or other extremes, as exemplified by MX missiles or superhardened siloes. In addition, a lowered priority for economies of scale would reduce the numbers of giant metallurgical or petrochemicals complexes, etc. Fewer technologies leading to monitoring tasks of mind-numbing boredom or requiring obsessive concentration would be evident. There would be some turning away from automation of interesting jobs.

As in many other scenarios for the future, communications and computers would be very important. What would differ would be how these systems could be used and who would have them; those differences would in turn imply different computer programs, different physical setups, and easier, more variable methods of operation.

The real changes would appear over time, as technology began to diverge more and more from the course it now is on. The new directions would be cumulative, successes building on successes, leading in turn towards a more democratic society, with citizens, groups, and communities more capable of acting to further the technological changes they wanted.

– Is this optimism warranted? What reason is there to believe the new technologies would be able to move us towards desirable social goals?

The main reason for believing that the program could succeed is that we know a great deal about other examples of technological change and the conditions for success. First, new technology programs are themselves nothing new. Governments or industries often decide on new courses to pursue and support these new projects with very substantial levels of funding; the projects often succeed or at least progress toward the desired goals. It is a general feature of technology that large increases of resources applied towards a new goal leads to rapid change, although within any one technology the changes often become slower after a while. The reason for this is simply that starting in a new direction involves a huge opening up of untried possibilities, and of these, quite a few turn out to work. If only one technology is pursued, then after a while all the easy ways to make progress in the desired direction have already been implemented, so change slows.

With the goals of this program, such as promoting democratic values and improving conditions for the worst off in society and in the world, even small changes would be of benefit. Enough smaller-scale technological efforts have been in accord with the values of this program to assure us that change in these directions is possible, and the smallness of the previous efforts also makes evident that possibilities are by no means exhausted. Numerous different technological paths can be tried, and the room for improvement is basically unlimited. So we have every reason to believe that a considerably stepped up effort to make changes in these directions would be more a success than a failure.

One major reason why many technological efforts often are less successful than anticipated is that they are aimed at obtaining a more or less permanent upper hand in some human conflict, which necessitates outthinking all possible future opponents. For example, burglar alarms, locks, anti-counterfeiting measures, as well as weapons and attempts to increase productivity to stay ahead of the competition, all ultimately fail because they are opposed by humans who can think beyond them, going further in imagination than the technologists who dreamt up these devices. The reinvented technology program as much as possible involves not attempting to outthink opponents, but attempting to increase access and capacities, that is, to extend possibilities rather than to foreclose them. Coopting these technologies for opposite purposes would mean finding ways to use them essentially as weapons, and while that cannot be precluded, there is no reason to suppose it would be especially easy to do. And, unlike attempts at technologies of social control – for that is what the more competitive efforts are – the thrust of the effort would be to reduce sources of conflict and resentment, not to increase them.

– *Assuming the political will existed, what would the changeover from current directions of technology to the new ones entail? How would such changes be possible? How would the new technological projects actually be planned?*
The process would have to be dialogic, which simply means that, through meetings, publications, hearings, conferences, and study groups the various parties (technologists,

representatives of workers, consumers, the poor, government officials, and others), there would emerge a sense of what specific technologies to pursue. A year or two of that dialogue would be ample for a rough start in this program, and, in fact, that sort of dialogue would very likely be part of the political effort needed if the program were ever to win anything approaching full Congressional acceptance. Since many of the individual projects would not be large-scale, there would be no particular difficulty getting them started, and for many, real progress would be quite rapid. (Of course, as work proceeded, it would become evident that technologies not previously included would be fruitful to pursue. At the same time, some area might turn out to be less promising than it seemed at first.)

From that point, the path to having the new items and processes available for use is less clear, but the initial political enthusiasm would probably work in favor of those steps going quickly as well. Those initial changes, if they worked at all, would not only help bring about support for subsequent changes, but also help strengthen those who had most reason to be supportive.

The projects are likely therefore not only to succeed, but to be very rewarding for those working on them. That leads to a further question.

– What would these changes mean for technologists?
Wouldn't some suffer as previous projects were cancelled?
It is true that some technologists would suffer as research programs such as the "Star Wars" efforts were stopped, since many of the special fields involved would not transfer easily to more humane applications. In particular, young technologists on the verge of making their reputations would lose much momentum in their careers. In general, however, technologists would not fare as badly as do ordinary workers, who today often face technologically induced unemployment with less education in adaptability and fewer options for change. Technological innovation projects frequently reach completion after a few years, whereupon the technol-

ogists involved must move on to something else. Moreover, the engineering profession has long been an insecure one, with older engineers often replaced by younger ones who work at lower salaries and have more up-to-date training. Despite such precedents, a program based on humane values should be consistent enough not to scorn technologists. Without going so far as to keep undesirable projects alive for that purpose, it should offer displaced technologists reasonable chances to decide how to redirect their skills in ways both socially useful and personally satisfying.

More broadly, technologists and eventually scientists would spend more of their time finding out about people's wants, not just for new products, but for ways of living. There would be a shift, at first subtle and then more prominent, in what it means to be a technologist, or, to put it another way, just what the relationship is between technologists and other people. This would change relations among technologists as well.

New people would want to enter technological professions, and these professions as well would have to change. Since the educational process for the profession, if it is to make sense, prepares the student for the life she or he will be leading as a professional, the content of courses and curricula will have to change too. One important aspect of professional education is that it not only prepares the student but inculcates a definite set of values along with preparation for practical work. These have to be linked. There's no point having a medical student solemnly take the Hippocratic oath unless the student sees the teachers practicing medicine accordingly. Furthermore, the student has to believe that, in the world of medicine she or he is about to enter, upholding the oath will really be expected – or, at least, practical.

The same would hold in technology. Changing the values in the curricula would be an important element of preparing students for their new roles in society, but for them to make sense, the change in values would have to come with the real changes of conditions of the profession. It wouldn't help for one course to emphasize that people must always come first, if in most others the criteria for design taught through practical problems (say, in an industrial setting) were least monetary cost and maximum efficiency.

Pouring some money into altering educational practices at the appropriate moment would help speed the entire program. So would special efforts to recruit new people and new types of people into the technological professions. The new types would include not only previously underrepresented social groups, such as women or blacks, but anyone with more human-centered outlooks who previously would have found technology unappealing.

Related professions would appear as well. One would involve an interpretive role: helping explain to technologists what the needs of particular groups – for instance, the elderly in a certain community – translate into in technological terms. This profession would both evolve out of and shade into the older technological professions, but would take on a new importance and demand a new combination of skills and knowledge. Industrial management currently shares some functions with this new profession. What differs is again a matter of emphasis and degree, since managers are understood to have a superior power position, and since for them the well-being of customers or workers is a secondary, rather than primary goal.

Yet another probable new profession would be more evaluative – examining the social implications of new technologies as they emerged, systematizing and extending analyses of the type found in this book. Among other places, the profession would be needed for the evaluative work mentioned in Chapter 11.

– *Returning to the overall program, how could it really be administered, especially at the beginning? How would the commissions represent the diversity of interests the programs presuppose? Wouldn't they be either at the mercy of a President who chose to stack them with supporters of one particular interest, or – through their members' lack of enough technological background – at the mercy of the technologists they would supposedly be directing? And if the new interests did exert control, wouldn't that be at the expense of sound technological judgement?*

The Reagan administration has demonstrated that current methods of appointing commissions do give too much power to whoever wins a majority of the electoral college. New methods of delegating nominating powers to groups who have more claim to represent the interests that various commissions are supposed to be serving are clearly needed. While no rule can be foolproof, given political support for such a change, it ought to be possible to devise a superior, workable system for selecting members for government commissions, including the ones proposed in this work.

As mentioned in Chapter 7, balancing sound technological judgement with the needs of those previously unrepresented in technological decisions does pose some challenge. But there are presumably enough representatives of groups such as unions or even welfare rights organizations who know something about technology, and enough technologists whose sympathies would coincide well enough with the values of this program, to make a start.

To propel the effort further, the national commissions could each encourage local boards. Through the elaboration of the program itself, these boards would begin to have members who could come closer to bridging the gap between the needs of the most disempowered and a detailed grasp of technological possibilities.

– Wouldn't all these commissions, agencies, and local boards add too much to the weight of government? Wouldn't they be unwieldy and also promote too large a bureaucracy?

As argued in Chapter 2, however they are made, technological decisions have the force of government. Furthermore, these decisions are usually arrived at through some sort of process of give and take, combined with administrative actions and formal and informal meetings of some group. The reason to have all these boards and commissions is to open up and democratize the process. The commissions would assure some role to groups who now have little say: they would help broaden power rather than increase the total weight of dominant groups on the lives of average people. All these bureaucracies would replace much the same sort of operations that now function with little accountability or democratic involvement.

No complex society could survive without some kind of structure for agreements, for making decisions, and for acting in association. The total structure might as well be thought of as the state. In a true democracy, everyone would be part of that structure and have an equal chance at a say. The state could only "wither away," in Marx's phrase, if government grew in such a way, and to such an extent, that everybody could take part. As it is now, some of the state in this sense is not considered part of the government, but is known as private enterprise, corporations, learned societies, etc. A different nomenclature does not in itself change power imbalances or alter oppression.

One more aspect of the administration of technological projects seems worthy of mention. Some agency or group of agencies would have to be responsible for deciding, in advance of immediate call for them, what fundamental technologies would likely be of enough use for particular goals as to deserve development. However this is done, there are three criteria that ought to be kept in mind.

1 One result of the current diversity and speed of technological development is that any kind of project which requires a long effort before it can be used for the purpose motivating it, and which is tied to a definite technology, is likely to be overtaken and outmoded by other ways to achieve the same effect. This applies to railway tunnels in Japan (outmoded by air travel), to nuclear power reactors in the United States (largely outmoded by lessened increases in power consumption due to conservation, and by smaller alternative power sources), and probably to projects such as controlled nuclear fusion development or large petrochemical complexes. And, largely owing to the enormous

versatility embodied in microelectronics and computers, it is likely that the time-scale of projects for which this issue is important will shorten. In administering any broad technological program, this possibility must receive careful attention. That might involve developing what might be called "metatechnology" – a way of deciding the optimum amount of time to delay projects, in case a superior alternative is developed. A ten-year project ought to be delayed, if, there is reason to believe that within a year of its start, it is likely to be outmoded by the invention of a means to achieve the same social effect in half the time.

2 "Technological pluralism" would mean that for any worthwhile goal a variety of different technical options should be encouraged. This would help permit flexibility for specific circumstances, as well as avoiding unanticipated, undesirable side effects.

3 In any situation when failures could cause serious problems, it is important not to rely on technologies of too great complexity, since with them there can be no reliable way to predict, and therefore to avoid, major accidents.

Science policy changes

– So far, science, as opposed to technology, has received brief mention. What would the effect of this program be on science, specifically on federally funded science?
Science and technology policy are often treated as one; while there are good reasons to do so, the distinction is not entirely to be neglected. This book focusses on technology, not science, and some questions relating to science policy would deserve separate treatment. But the programs in Part Two do have two major kinds of implications for science policy.

First, since what is scientifically interesting is partly determined by technological needs, changes in the direction of technology would explicitly affect science; that was at least mentioned in relation to the Social Goals-Directed Research and Development Policies (Chapter 8), the Public Technologies Programs (Chapter 9), the Intellectual Claims Act – among other means of influencing private sector innovation (Chapter 10), and the Peaceful Global Technological Cooperation Program (Chapter 12).

Redirection obviously implies curtailing some lines of research while expanding others. For instance, science connected specifically with weaponry, such as the chemistry and metallurgy of plutonium or the physics of high power lasers, would likely lose support. It is less easy to say what fields would gain, since that would depend on what knowledge turns out to be lacking for advancing the more socially beneficial technologies. In general, applied science would, of course, be more directly influenced by changes in technological direction, while basic science, more remote from applications, would presumably change in more subtle ways.

The changes in science, like technological changes, would be accompanied by changes in supporting institutions. For instance, the Defense Advanced Research Projects Agency, which funds a variety of scientific fields, would not just be renamed, since its structure helps perpetuate the wrong mix of sciences that have been supported for the wrong reasons. While there could be some sort of transition period, new agencies would be needed – agencies with more explicit ties to the Social Goals-Directed programs and the other programs suggested in Part Two.

With all these changes, scientists would find that, just as for technologists, the values informing their work had changed, as had the values that there would be reason to include in educating young scientists.

The second broad set of implications relates to the fact that science itself is an area of human activity in which technology is applied. Quite a few of the proposals in Part Two for technology in general would make sense if carried out for technologies of use in science. For instance, the idea of mass distribution of new technologies (Chapter 9) suggests a strategy for making "state of the art" scientific apparatus more widely available than is presently the case, thereby increasing the number of individual scientists and laboratories that could operate in the forefront of various fields.

Technology for widening cultural equality (Chapter 8) might be applied to make more broadly available both the results of science and the possibility of participating in at least some scientific work. The Universal Access Information Network would also help make scientific work more broadly accessible. This and other aspects of the whole program could greatly increase dialogue between scientists and the rest of society, which could turn out to be of value for both parties.

Is reinvented technology affordable?

– How much would all this cost, and in a time of great deficits, how could it be paid for?
To be worthwhile, these programs would have to be capable of spending amounts comparable to other major technological development programs, eventually possibly in the range of $10–50 billion a year, not including the public works programs of Chapter 9, which could cost much more – perhaps $100–200 billion annually. The moratorium on military research, development, and procurement described in Chapter 12, on the other hand, would save around the same amount.

The costs would largely represent a reallocation of resources within the government. In the case of public technologies, some would involve a reallocation from private to public, in effect a transfer of responsibility and control over certain technological choices. The sums involved may seem huge but they are being spent today as private spending on related items that are less valuable than what we could have.

The Universal Access Information Network, for instance, would be a public version of what is now being put together privately in bits and pieces, with no one taking responsibility for making the best, most versatile, or fairest system possible. The justification for paying so much has to be in terms of the social value of the redesign of the communications system that the transfer of authority would signify.

Whatever the actual costs, at present the giant federal deficits make it very difficult to propose new programs, especially non-defense ones. How do we overcome the enormous obstacle of the deficit? A good place to begin is by noting that the deficit is a social creation even less unchangeable than technology. It itself is the consequence of a complex set of assumptions and commitments, including commitments not to raise taxes, to spend on defense, and to pay interest to creditors at a certain rate. Those commitments are not irrevocable and not fixed in stone, and it would be folly to regard them as such, with no heed to the consequences.

More deeply, economics only utilizes the means of money to deal with the distribution of resources, human capabilities, social possibilities, and levels of desire. It should only be taken to be constraining when the particular counting system it has adopted is working well; if the program outlined here would involve better using resources and capabilities that exist, improving the qualities of people's lives and the cohesion of our society, then any accounting that claims it cannot be begun is simply nonsense.

At previous times of transition in American history, new financial means were invented to overcome similar traps. Creative financing is a daily reality on Wall Street, where "greenmail" and "junk bonds" have recently been perfected for far more dubious ends. Creative financing to get around the obstacle of the deficit could make much more sense and eventually will be necessary in any case, if current economic paradoxes are not to grow more severe. The question, then, is not whether the program can be afforded but whether it is a worthwhile way to use time and energies that we certainly possess.

14 Epilogue: How to use this book

This book has been an extended argument that the trend of
technology not only is extremely important, which almost
no one doubts, but that it is subject to human decisions.
And this is true in more than an abstract sense. It is possible
to change the policies of the US government in ways that
can make an enormous difference, not just for us in
America, but for the world.

What we would have if we make such changes would be
a new start, not only for technology, but for democracy, for
human equality, for a peaceful world, for richer cultural
and social possibilities, for a more stable and more just
economic system.

Needless to say, the argument is incomplete. Whole areas
of policy that it would make sense to change congruently
were hardly mentioned. And, in the areas covered, a wealth
of detail has been left out. Part Two has been in the form
of a program of policy changes. But it can make sense as a
program only if it is widely adopted or, in other words, if
it is widely considered politically possible. How could this
happen and who would be involved? And in the process,
what use can this book be?

The first way this book might be used is simply as an eye-
opener. For anyone who has chosen to read it, it will have
accomplished something if it convinces her or him that we
need to think more about the aims of the policies that shape
technology, and that an alternative policy can actually be

constructed. That level of acceptance would be a prod, if not to immediate action, then to reflection and discussion on the outlines of the policy.

The second way this book might be used would entail taking the program as stated as a valid point of departure, adopting this version as a framework for discussion, as a basis for possible political agreements – even as a nucleus, perhaps, of a new kind of political alliance encompassing all those (overlapping) constituencies that would benefit in specific ways: women, the poor, minorities, labor (both organized and not yet organized), the peace movement (both the antinuclear weapons wing and the anti-intervention wing), supporters of a just development process in the Third World, and the socially concerned wing of the environmental movement – all these, together with the technologists and scientists who want their work to benefit and not to harm present and future generations alike. It is an alliance easy and obvious to describe on paper, but it does not yet exist in practical terms.

Were such an alliance to emerge, what could it do? The first step, of course, would be for its members to talk, to try to understand more specifically each others' motivations, needs, and aims. The next step might be to short-circuit all that was said here about government policies, and to begin to develop some of the new ties in practical terms, some new technologies, some new sources of advice, perhaps some new legal strategies (such as extending the notion of malpractice, as was indicated in Chapter 11). Another possibility would be informal arrangements for sharing, or simply not bothering to register, intellectual property.

In pursuing those directions, what might actually be forming is a somewhat different kind of alliance, one that is already nascent without this book: all the efforts at alternative community-scale technologies; the conversion movement (for converting from a war to a peace economy); movements for promoting public domain software; and movements for workers' rights relative to technology. For these and related movements, the book might serve as a minor catalyst, perhaps slightly speeding up a process already in motion.

As the accomplishments of that alliance gain in visibility, they could work as illustrations of some of the points raised in this book. For the poor, for example, such a movement might illustrate how they have been deprived in the area of technological development, as well as in the more obvious ways, and how important that deprivation is. It may also illustrate another thesis. That is, worthy as they may be in themselves, informal, unofficial sorts of technological change are likely to be overwhelmed by the thrust of corporate and government development. Some way must be found to influence the latter processes more definitely. The informal alliance could, of course, become the core of a movement for promoting changed government policies.

This country is fortunate in having at least three levels of government with some degree of sovereignty – federal, state, and local. Because technology is universalizable, some activities at even the lowest level can have wide influence. In addition, some of the larger universities and medium-size corporations sensing new market niches could also begin to develop significant technologies, possibly inspired in part by this book, on scales larger than informal groupings might be able to accomplish. All that could begin to make a movement with a real and effective presence in the world.

The federal government is also not a monolith. Laws and programs can often be changed incrementally. It is an unfortunate reality that many legislative initiatives supporting crude extensions of intellectual property, new military technologies, increases in uncritical support for automation or other aspects of high technology, all often pass into law with little organized opposition. If, as a result of this book, that process becomes just a little harder, that would be of great value.

Better yet, this book, and the movements already mentioned, might begin to influence the formation of new legislation, quite possibly unambitious at first, that would tend in the direction of a reinvented technology policy.

Eventually, the whole program might appear on the legislative calendar in some form, perhaps at first under the
sponsorship of only one or two Senators or Representatives,
who would clearly be intending no more, at that stage, than
a symbolic gesture.

Is further motion possible? Passage of a full program,
resembling what has been outlined in Part Two as a whole,
of course, appears currently as a faint and distant prospect
at best. The various movements and groups that are its
most logical supporters are either being partially coopted by
programs such as "Star Wars", or are reeling under a series
of economic and government blows, some of which are
described in Part One.

If thinking about new technology involves one step of
projection into the future, thinking about policies for
redirecting technology involves two such steps. This cannot
be easy for movements and individuals who are almost daily
on the defensive. Yet the alliances just discussed could form
and could grow; and the tide of events could make ever
more evident why a decisive switch in course is needed.

More than once, the passage of twenty years or so has
turned fringe ideas into unquestioned commonplaces. That
is a lot to ask for any idea. Yet it is difficult to be sanguine
with such a thought. In the onrush of events, twenty years
can be a very long time to fail to change course. There may
be no way to prevent technologies from getting worse before
they get better. And that implies that there is no way to
avoid intensified human misery, for this program or something else with much the same force would be needed to do
that.

One last point: it may seem that this analysis accepts only
two poles – unquenchable optimism or darkest pessimism.
Is a middle ground possible? And is that not what we are
likely to muddle towards?

Whereas humanity was able to muddle through in the
past, in fact that meant great suffering for many. A social
order that turns its back on preventable suffering of great
magnitude is in serious moral straits. How different is it

from the "good Germans" who took care not to notice the slave labor and concentration camps all around them? While societies of the past might have been able to survive that level of moral blindness, with technologies of destruction so well developed, fear of revenge by the victims, if nothing else, should indicate that those options no longer suffice. A decent life for all or vast destruction may be the only real choices.

It would be wrong to view the alternatives offered here as merely utopian. In large measure, they are means to spread the capacity for muddling through to far more of the world than has it now. In a crucial sense, that capacity is the single prerequisite for any feasible good life.

The only real grounds for optimism come in refusing to accept a prepackaged fate – in this case, refusing to accept the fatalistic notion that only one technological future is conceivable. Reinventing technology is therefore not so much a completable program, as a part of what human life will continue to be about, if life is to continue.

Endnotes

A brief introduction to general sources for the study of technology, its social origins and impacts

Direct observation, even of ordinary daily life, reveals much of the nature of technology and its place in our society. This can be supplemented through a virtually inexhaustible wealth of materials. One place to begin is a good daily newspaper, such as *The New York Times*. Further sources presenting greater detail include popular and semipopular science and technology journals, from *Popular Science* or *Popular Electronics*, through various computing magazines, to *High Technology*, to the British *New Scientist*, to *Scientific American*. More scholarly, professional or industry-oriented journals abound; many cover specialized technologies, particular industries or crafts.

To obtain a sense of the breadth and range of technologies, it is useful to peruse compilations such as *Thomas's Register of American Manufacturers* – a classified catalogue of industrial products – or course catalogues of engineering and technical schools and departments. Numerous government agencies produce publications offering various glimpses of technology. The reports of the (Congressional) Office of Technology Assessment (OTA) and of the National Research Council (a branch of the National Academy of Sciences) are often concerned with the social impacts of technological choices. (See Chapter 5 for some comments on these.) Other treatments of various issues involving the interaction of technology and society are to be found in legal journals; topics of interest include intellectual property, labor relations, corporate law, and malpractice.

As with most other areas of life, other sources are histories, memoirs, and biographies. Also, there are many more specifically focused books for general audiences. The reviews of these works are an important additional source. Imaginative works encompassing fiction (including, but not limited to science fiction), the visual arts, movies, radio and television productions of course reflect and illuminate the broader society and its values in ways that often elude more analytic approaches; frequently, and often unexpectedly, these arts offer insights into technology.

A number of scholarly and semi-scholarly journals also cover some aspect of technology's social or political impacts, and technology policy, etc. A new, already influential, and quite revealing journal is *Issues in Sciences and Technology* a publication of the National Academy of science intended for "the leaders of the 1980s" and leaving unquestioned such mainstream assumptions as the desirability of economic growth or preparedness for war.

Journals with definite and explicit sets of values focussing on technology – especially from a liberal, left-liberal, or radical perspective are virtually non-existent; exceptions, such as *Science for the People* and the *British Radical Science Journal*, tend to have small budgets that not only limit their coverage but sometimes have led them to favor overly simplistic analyses. The news sections of both *Science* and *New Scientist* generally have a moderately liberal cast, but the shortness of items tends to be a drawback. The *Bulletin of the Atomic Scientists* is a good example of a generally liberal journal that focuses on a narrower range of issues: primarily the arms race and the proliferation of nuclear technologies. The United Nations journal *Development Forum* often represents the viewpoint of those who support independent development of and equality within Third World nations. A number of books with perspectives related to these are included in the Select Bibliography below.

Notes and References

The phrases following the page numbers identify the annotated passages. Citations in these notes refer to items in the Select Bibliography.

p. 8   *Broadly diffuse concept.* Many philosophically or sociologically intended discussions of technology become incoherent because insufficient regard is paid to nuances of definition. One source of confusion is that the English words "technology" and "technique" do not correspond exactly to their French or German counterparts (the same is true for "science.") Recent debates have been international, so that the concepts get mistranslated back and forth many times over. In using them here, I have attempted to be faithful to common American usages, while at the same time exploiting connotations that it seems to me the usages carry with them.

p. 11 *Technology and modernity*. Anthropologists, historians, and others have written as if every society is a technological one, so that while technologies may vary, technology is timeless; they may speak, e.g., of "neolithic (new stone-age) technologies." In addition, some anthropologists suggest that singling out modern "Western" practices only as technology is ethnocentric. Clearly, no pure "pretechnological" or – better – non-technological society is available for our direct scrutiny. Conclusions about the nature of such societies and innovation within them are necessarily somewhat speculative and easily distorted by preconceptions; yet the distinctions offered in the text seem to do honor to available evidence without being ethnocentric; "technology" simply denotes a set of practices and social relations that are not found in other cultures. Nothing pejorative need by implied by keeping such distinctions.

p. 12 *Autonomy of technology*. Both critics and ardent supporters of technology have held it to be autonomous, and value free, a force acting on its own. Jacques Ellul is probably the most sophisticated and articulate critic of technology who upholds this view. While much in the presentation of technology offered here is in accord with Ellul, I do not agree with him in either his overly wide definition of technology or on this important point. "Technology," he says, "escapes any system of values" (Ellul, 1980). He also cites the example of technologically induced speed as having become an end in itself. In fact technology has accentuated the social valuing of speed but the underlying value came first (as captured in the legal phrase "time is of the essence"). The fact that technologies of slowness, for instance, are less highly developed and therefore less available has everything to do with dominant values. When these values change, as did happen to some extent in the US in the 1970s, fascination with speed can decline, and technology can change course.

Critics of weapons technology, including feminists, and left-wing critics of the Soviet Union's copying of American high-efficiency production techniques (e.g., Braverman) are among the sources of the views I have tried to spell out in the text.

p. 20 *Lakatos*, esp. pp. 86–9, presented his ideas in the context of a debate that also involved Thomas Kuhn, Karl Popper, and Paul Feyerabend. Despite obvious disagreements, all these authors seem to agree that the truth of science is at best a matter for social consensus, and is not subject to any rigorous proof. They all seem to be weak when it comes to considering the relation between the scientific community and the wider society. Lakatos himself does not specify that the new experiments and concepts have to be interesting, but that seems to be implied by his argument.

p. 32 *Specific impacts of technology*. See, for example, Gerbner on the impact of television, and see Hirschhorn, Noble (1984), and Shaiken on different impacts of factory automation.

p. 34 Landes describes changes in the watch industry.

p. 40 *Work and technology*. There is an immense literature on the relationship between technology, work, and workers. See, for example, Aronowitz, Blumenthal, Braverman, Burawoy, Leontief and Duchin, Goldhaber (1983), Hirschhorn, Noble, Melman. Additional sources would include works from the fields of industrial psychology and sociology, industrial engineering, industrial management studies, histories of the labor movement, etc.

p. 42 *Housework*. For critiques of the commonly accepted notion that "labor-saving" technologies reduce housework, see, e.g., Cowan, Illich (1983), Thrall. As far as I am aware, the general impact of technology on cultural practices within the home has not been studied much, although certainly noted in fiction, movies, etc.

p. 44 *Elimination of farm labor*. See Martin for a conscious plan for this approach.

p. 47   For a discussion of *comparative advantage*, see Zysman and Cohen.

p. 51   *Kondratieff cycles*. See, e.g., Marchetti.

p. 65   *Television and violence*. The best work in this area is Gerbner.

p. 73   *Interesting jobs*. For a clear account of some of these jobs in automated settings see Hirschhorn.

p. 88   *Exact opposite effect*. Barnet argues this well.

p. 101   Blumenthal, *et al. OTA*. At its best, OTA's work can be excellent, while not evading the problems indicated. For example, see its study of programmable automation.

p. 109   Examples of conservative views of technology are to be found in Gingrich and Ramo as well as Simon and Kahn.

p. 112   *Duplicating Silicon Valleys*. For instance, see the views of Babbitt, Governor of Arizona.

p. 113   *Industrial policy*. See Reich.

p. 117   *"Better is better"*. see Hart and Fallows. Fallows is typical of new thinkers on defense in starting from the premise that the United States will always have undefined "interests" that must be defended. By ignoring the issue of what such interests might be, this premise supports an open-ended growth of military technologies (even if shaped according to Fallows' criteria) and totally discounts the possibility that world peace might genuinely be attainable.

p. 147   "Female"-valued technology. See, e.g., Haraway and Keller.

p. 148   *Unemployment*. Hacker argues this level of unemployment.

p. 161   *Childhood*. See Postman for further arguments on this point.

p. 182   See *Industrial* . . . for broad discussion of current problems with intellectual property laws.

p. 209   *Workers' rights*. See International Association of Machinists and Aerospace Workers for a workers' bill of rights. That material derives in part from existing rights in Norway and elsewhere in Europe.

p. 219   *Global technological cooperation*. See Independent Commission (the Brandt report) for a call for what in part is a similar approach.

Select bibliography

Aronowitz, Stanley, "Why Work?" *Social-Text*, 12, pp. 19–42, Fall, 1985.

Babbitt, Bruce, "Grassroots Industrial Policy," *Issues in Science and Technology*, vol. 1, no. 1, pp. 84–93, Fall, 1983.

Barnet, Richard J., *Real Security*, New York, Simon and Schuster, 1981.

Bell, Daniel, *The Coming of Post-Industrial Society*, New York, Basic, 1973.

Bernal, J. D., *Science in History*, 3rd edition, Cambridge, Massachusetts, MIT Press, 1971.

Blumenthal, Marjory S. *et al.*, *Computerized Manufacturing Automation: Employment, Education and the Workplace*, Washington, DC, US Congress, Office of Technology Assessment, OTA-CIT 235, April, 1984.

Braverman, Harry, *Labor and Monopoly Capital*, New York, Monthly Review, 1974.

Bronowski, Jacob, *Science and Human Values*, New York, Harper & Row, 1975.

Burawoy, Michael, *Manufacturing Consent*, University of Chicago Press, 1979.

Burnham, David, *The Rise of the Computer State*, New York, Random House, 1983.

Cowan, Ruth Schwartz, *More Work for Mother: The Ironies of Household Technology from the Open Hearth to the Microwave*, New York, Basic Books, 1983.

Dreyfus, Hubert, *What Computers Can't Do: The Limits of Artificial Intelligence*, New York, Harper & Row, 1972.

Eco, Umberto, *A Theory of Semiotics*, Bloomington, Indiana University Press, 1976.

Ellul, Jacques, *The Technological Society*, New York, Knopf, 1964.

Ellul, Jacques, *The Technological System*, New York, Continuum, 1980.

Ernst, Dieter, *Restructuring World Industry in a Period of Crisis – the Role of Innovation: An Analysis of Recent Developments in the Semiconductor Industry*, Vienna, United Nations Industrial Development Organization, 1981.

Fallows, James, *National Defense*, New York, Random House, 1981.

Ferkiss, Victor C., *Technological Man*, New York, Braziller, 1969.

Feyerabend, Paul, *Against Method*, London, New Left Books, 1975.

Florman, Samuel C., *Blaming Technology*, New York, St. Martin's Press, 1981.

Foucault, Michel, *Archaeology of Knowledge*, New York, Harper & Row, 1976.

Friedrichs, Guenter, and Schaff, Adam (eds), *Microelectronics and Society: A Report to the Club of Rome*, New York, New American Library, 1982.

Gerbner, George, "The Mainstreaming of America: Violence Profile Number 11," *Journal of Communications*, Philadelphia, Annenberg School of Communications, University of Pennsylvania, Summer 1980.

Gingrich, Newt, *Window of Opportunity*, New York, St Martin's Press, 1984.

Goldhaber, Michael, "Politics and Technology: Microelectronics and the Prospects of A New Industrial Revolution," *Socialist Review*, 52, pp. 9–32, Oakland Ca., New Fronts Publishing, July–August, 1980.

Goldhaber, Michael, "Microelectronic Networks: A New Workers' Culture in Formation?," *Critical Communications Review*, Vincent Mosco and Janet Wasco (eds), vol. 1, Chapter 10, pp. 211–43, Norwood, New Jersey, Ablex, 1983.

Goldsen, Rose K., *The Show and Tell Machine*, New York, Dial Press, 1977.

Gorz, André, *Goodbye to the Working Class*, Boston, Beacon, 1983.

Greenbaum, Joan M., *In the Name of Efficiency*, Philadelphia, Temple University Press, 1979.

Greenberg, Daniel, *The Politics of Pure Science*, New York, New American Library, 1967.

Habermas, Jürgen, *Knowledge and Human Interests*, Boston, Beacon, 1971.

Hacker, Andrew, "Where Have the Jobs Gone?" *New York Review of Books*, vol. XXX, 11, pp. 27–31, June 30, 1983.

Haraway, Donna, "A Manifesto for Cyborgs: Science, Technology and Socialist Feminism in the 1980s," *Socialist Review*, no. 80 pp. 65–101, Center for Social Research and Education, Berkeley, California, 1985.

Hart, Gary, *A New Democracy*, New York, Quill, 1983.

Hirschhorn, Larry, *Beyond Mechanization*, Cambridge, Mass., MIT Press, 1984.

Hofstadter, Douglas, *Metamagical Themas*, New York, Basic, 1984.

IEEE *Spectrum*, New York, Institute of Electrical and Electronic Engineers, vol. 81, no. 3, *passim*, June, 1984.

Illich, Ivan, *Tools for Conviviality*, New York, Pantheon, 1975.

Illich, Ivan, *Gender*, New York, Pantheon, 1983.

Independent Commission of International Development Issues (Brandt Commission) *North–South, A Programme for Survival*, Cambridge, Mass., MIT Press, 1980.

Industrial Innovation and Patent and Copyright Law Amendments, Hearings of the Subcommittee on Courts, Civil Liberties, and the Administration of Justice of the House Committee on the Judiciary, April 3, 15, 17, 22, 24, May 8, and June 9, 1980, Washington DC, US Government Printing Office, 1980.

International Association of Machinists and Aerospace Workers, *Let's Rebuild America*, Washington DC, 1983.

Jones, Barry, *Sleepers, Wake! Technology and the Future of Work*, Melbourne, Oxford University Press, 1982.

Keller, Evelyn Fox, *Reflections on Gender and Science*, New Haven, Yale University Press, 1985.

Kraft, Philip, *Programmers and Managers*, New York, Springer-Verlag, 1977.

Kuhn, Thomas S., *The Structure of Scientific Revolutions*, University of Chicago Press, 1962.

Lakatos, Imre, *The Methodology of Scientific Research Programmes*, Cambridge University Press, 1978.

Landes, David S., *Revolution in Time*, Cambridge, Mass., Belknap, Harvard, 1983.

Leontief, Wassily, and Duchin, Faye, *The Impacts of Automation on Employment, 1963–2000*, New York, Institute for Economic Analysis, New York University, 1984.

Levy, Steven, *Hackers*, Garden City, NY, Anchor Press/Doubleday, 1984.

Marchetti, Cesare, "The Automobile in a System Context: The Past Eighty Years and the Next Twenty Years," *Technological Forecasting and Social Change*, vol. 23, pp. 3–23, March 1983.

Marcuse, Herbert, *One-Dimensional Man*, London, Routledge & Kegan Paul, 1964.

Martin, Phillip L., "Labor Intensive Agriculture," *Scientific American*, vol. 249, no. 4, pp. 54–9, October, 1983.

Marx, Karl, *Capital*, vol. I., London, Penguin, 1976.

Melman, Seymour, *Pentagon Capitalism*, New York, McGraw-Hill, 1970.

Nelkin, Dorothy, *The University and Military Research*, Ithaca, Cornell University Press, 1972.

Noble, David, *America by Design*, New York, Knopf, 1977.

Noble, David, *Forces of Production*, New York, Knopf, 1984.

Norman, Colin, *The God That Limps*, New York, Norton, 1981.

O'Neill, Gerald K., *The Technology Edge*, New York, Simon & Schuster, 1983.

Popper, Karl, *The Logic of Scientific Discovery*, New York, Harper & Row, 1959.

Porat, Marc Uri, *The Information Economy*, Washington DC, US Department of Commerce, 1977.

Postman, Neil, *The Disappearance of Childhood*, New York, Delacorte, 1982.

Ramo, Simon, *America's Technology Slip*, New York, Wiley, 1980.

Reich, Robert, *The Next American Frontier*, New York, Times Books, 1983.

Roos, Patrica A., *Gender and Work: A Comparative Analysis of Industrial Societies*, Albany, New York, SUNY, 1985.

Roy, Rustum, *Experimenting with Truth*, Oxford, Pergamon, 1981.

Roy, Rustum and Shapley, Deborah, *Lost at the Frontier*, Philadelphia, ISI Press, 1985.

Rybczynski, Witold, *Taming the Tiger*, New York, Viking, 1983.

Shaiken, Harley, *Work Transformed*, New York, Holt, Rinehart & Winston, 1985.

Simon, Julian L. and Kahn, Herman, (eds) *The Resourceful Earth: A Response to Global 2000*, Oxford, Blackwell, 1984.

Smith, Alice Kimball, *A Peril and a Hope: The Scientists' Movement for the Control of Atomic Energy*, Cambridge, Mass., MIT Press, 1971.

Sohn-Rethel, Alfred, *Intellectual and Manual Labor*, Atlantic Highlands, NJ, Humanities Press, 1978.

Thrall, Charles M., "The Conservative Use of Modern Household Technology," *Culture and Technology*, University of Chicago Press, vol. 23, no. 2, pp. 175–94, 1982.

Toffler, Alvin, *The Third Wave*, New York, Morrow, 1980.

Toffler, Alvin, *Previews and Premises*, New York, Morrow, 1983.

Turkle, Sherry, *The Second Self*, New York, Simon & Schuster, 1984.

Weizenbaum, Joseph, *Computer Power and Human Reason*, Freeman, San Francisco, 1976.

White, Lynn, Jr., *Medieval Technology and Social Change*, Oxford University Press, 1962.

Winner, Langdon, "Do Artifacts Have Politics?," *Daedalus*, Winter, 1980.

Zysman, John and Cohen, Stephen S., "The Mercantilist Challenge to the International Trade Order," report presented to the Joint Economic Committee, US Congress, November, 1982.

Index

statements, 205; and
intellectual property, 195–6
Software: 57, 154; and ethnicity
and gender, 75; nonproprietary
market for, 164–5; pace of
development of, and social
power, 65; and weapons, 65–6
*Special Groups, Agency for
Technology for,* 159–60
"Star Wars": 65; and the right,
110–11; as trend, 81
State, "withering away," 235
Status: *see* Hierarchy
Strategic Defense Initiative; *see*
"Star Wars"
Structural transformation, 51
Sunrise industry metaphor, 3, 4
Surprise, technological,
preventing, 92–3

Tax policy, 100, 199–20
Technological pluralism, 236
Technologists: defined, 9;
educational process and values,
232–3; how affected by new
program, 233; stake in
appearing apolitical, 104; and
technology, 9
Technology, defined, 8–12; nature
of reinvented, 228; transfer of,
50–1
Third World, development of: and
gender, 43; and industrial
policy, 116; and intellectual
property, 192; and
technological cooperation, 220;
tragedy of, 137
Trade secrets: 99; *reduction of,*
189, 193
Tragedy of development, 137
Transfer of technology, 50–1; *see*

also global, Third World,
villages
Transition: times of, 51–4; to new
policies, 230–1
Transportation, 6
Truth, scientific, 21

*Universal Access Information
Network:* 172–6; and
Intellectual Claims, 193–4
Universalizability of technology,
25–8, 47–51

Values: balanced, 170: democratic
5, 123–38; inconsistencies of,
25; of military, 89; necessarily
guiding technology, 15; non-
dominant, and technology, 25
Villages, *Agency for Technologies
for Aiding Development of,*
169; technologies for, 137
Violence and media, 69–71

War, foreseeing, 226
Wealth, redistribution of, and
technology, 33–6, and high
technology, 61–3; two aspects
of, 33–4
Weapons technology: *see*
Defense, Department of:
electronic battlefield; power;
smart weapons; and Star Wars
Work: *agencies relating to,*
143–54; definition of, 149,
Improved quality, Office for,
144–6; and technology, 40–1;
two meanings, 103–4
Work week, shortened, 152
World order, 225–6